HANDBOOK OF SUSTAINABLE TEXTILE PRODUCTION

Handbook of Sustainable Textile Production

Editor

Abdulkerim Macar

scitus academics

Handbook of Sustainable Textile Production
Edited by **Abdulkerim Macar**

ISBN: 978-1-68117-576-8
Library of Congress Control Number: 2016943011

SCITUS Academics LLC,
www.scitusacademics.com
Box No. 4766, 616 Corporate Way,
Suite 2, Valley Cottage,
NY 10989

Notice

Printed in United States of America on Acid Free Paper

PREFACE

Textile products are manufactured, distributed, sold and used throughout the world. Therefore, a quantitative evaluation of the sustainability of the textile production chain is very important. Sustainable textiles refer to materials derived from environmentally friendly resources such as fiber crops grown sustainably or possibly recycled. In addition, it refers to how these clothes are made. Production considerations include the water and energy used in production, the impact of waste in production and the social responsibility of a company towards its employees and the communities around its production facilities. In addition to environmental and health risks, the addition of chemical finishes to textiles may adversely affect the sustainability of fabrics at the end of their lives. Natural materials such as wool or cotton, which could biodegrade after use, cannot do this safely if they are loaded with chemicals. Added chemistries, including coloring, coating and finishing, can affect the overall health of textile workers and end product customers.

The book offers the possibility of a modular analysis of all relevant areas, including quality management, use of chemicals, environmental management, environmental protection, public responsibility and safety and health. This book is a compilation of technical, economic and environmental information from the different processes in this chain. The book highlights and evaluates the environmental and social impact of apparel. It examines the complexities associated with the implementation of sustainable measures in the massive supply chain of apparel.

TABLE OF CONTENTS

1

SUSTAINABLE APPAREL? IS THE INNOVATION IN THE BUSINESS MODEL? - THE CASE OF IOU PROJECT

Miguel Angel Gardetti[1*] and Subramanian Senthilkannan Muthu[2]

[1]Sustainable Textile Center, Av. San Isidro 4166 PB A, C1429ADP Buenos Aires, Argentina.

[2] Sustainability Consultant, SGS (HK) Limited, Global Sustainability Services, Hong Kong.

ABSTRACT

Since sustainability approaches to the apparel industry are based on incremental rather than transformative solutions, the prevailing business models need to be changed.

This paper will introduce the IOU Project as an example about disruptive business model in the apparel and fashion sectors. This case was selected since disruptive innovation requires an entrepreneurial mindset and spirit with close coordination and cooperation among all participants (JAMA 352:24-28, 2004). The analysis - which is based on the Clark, Osterwalder, and Pigneur's (2012) business model (personal) 'canvas' - puts special emphasis on the supply chain development and on the relationship developed between the initiative and its customers, as well as on whether or not it fostered a relationship between the customer and the artisans who were part of the supply chain.

To develop this case, the authors collected background and qualitative and quantitative information of the company from its participation in the Best Performance in Sustainable Luxury in Latin America Award, getting this award in the category 'Best Fashion and Accessories Enterprise of Sustainable Luxury' in the 2011 edition. This source was supplemented with the chapter written by Kavita Parmar (2014) in the book entitled Sustainable Luxury and Social Entrepreneurship - Stories from the Pioneers, whose editors were Miguel Angel Gardetti and María Eugenia Girón (2014). Moreover, the authors considered the e-mails exchanged with Ms. Parmar within the framework of the above award.

Using a disruptive business model, the IOU Project manages to develop a value chain based on craftsmanship. Craftsmanship breathes humanity into objects, allowing the individual to create for the individual. In doing so, it translates the principles of knowledge, purpose, and timelessness into products that represent unique achievements (Ricca and Robins, 2012:53). Within this framework, supply chain management became a strong driver of value, for the IOU but for artisans and customers as well (Kell,2010).

Finally, the IOU business model helps develop a strong relationship with consumers and artisans which dignifies the latter and appreciates their work.

KEYWORDS

Social impacts, Environmental impacts, Supply chain, Disruptive, Sustainability, Innovation, Business models

1. BACKGROUND

Any literature related to sustainable development refers or begins with the definition of sustainable development, which was defined as, 'development that meets the needs of the present without compromising the ability of future generations to meet their own needs' in Our Common Future, also known as the Brundtland Report, which was written in 1987 (World Commission on Environment and Development1987). This definition has undergone many critiques, as there were not many issues in the 80's which we are facing now; and we must say that if this report were written today, sustainable development would most likely be defined in a different way to portray the current situation. To define what sustainable development is all about is quite a common, though difficult, issue (Sneddon and Howarth 2006) and from the entrepreneurial point of view the elements under the umbrella of sustainability - pollution prevention, resource productivity, environmental management, sustainable technology, social entrepreneurship, corporate social responsibility, human and labor rights, transparency, stakeholder management, among others - are increasing day by day.

This same ambiguity of defining sustainable development prevails in the clothing and apparel sector, and it becomes more difficult to define what a sustainable apparel/clothing product is about, especially when it comes to the apparel sector. In the global economy, the role played by the world's garment and textile industry is significant, and the global turnover of this industry in 2011 was almost USD 3 trillion (including clothing, textiles, footwear, and luxury goods). This sector has significant levels of environmental and social impact, and it has not yet developed an overall vision to achieve sustainability up to the mark (Maximilian Martin, 2013; Muthu, 2014). The textiles and apparel sector comprises a lengthy supply chain, which is subject to mass environmental and social impacts.

Most of the current innovations in the apparel industry intended to achieve sustainability focus on raw materials (that is to say, eco-friendly raw materials) and products. Nevertheless, innovations are also being carried out in the process, logistics, recycling, and upcycling. Along this line, the World Economic Forum (2010) summarizes the limitations of current sustainability initiatives as follows:

current trends towards sustainability are welcome but inefficient. The shifts ... are focused on incremental, rather than transformative, change.

While they may improve sustainability at the margins, they are rooted in a model of consumption that is itself unsustainable.

But very few innovations only focus on the total business model, which is the havoc of sustainable innovations of the current apparel sector.

There is a need to redefine dominant approaches to business in the apparel sector. Most industries are still characterized by a 'dig it up-use it-throw it away' production system which prevents the achievement of sustainable development in any meaningful way (Wells 2008: 85). Decades of talk about corporate social responsibility (CSR), corporate citizenship, corporate sustainability, and other related buzzwords have not radically changed this picture.

A transformation of dominant business models is therefore needed in order to achieve systemic changes toward sustainability. Rather than perceiving sustainability solely as compliance and risk management, businesses have to recognize the value-creating potential of sustainable business models. However, so far, literature that bridges business model and sustainability thinking is relatively scarce. Stubbs and Cocklin (2008: 103) argue that 'understanding of sustainable business models and how sustainable development is operationalized in firms is weak', and Schaltegger et al. (2011: 12) argues that 'Neither theoretical nor empirical research offers sufficient answers to the question what a sustainable business model might be'.

This work explores the potentials for building business models for sustainability through innovation or radical innovation through the analysis of the case called the IOU Project, created by the young entrepreneur Kavita Parmar, an e-commerce initiative that offers fashion-conscious consumers a very distinct platform. Clothes sold on the Web are designed and manufactured by European artisans. These clothes are based on hand-woven fabric typical of India -*madras*- made by Indian artisan knitters. Kavita Parmar's project was selected for the analysis since the disruptive innovation requires an entrepreneurial mindset and spirit with close coordination and cooperation among all participants (Rainey, 2004).

2. METHODS

To develop this case, the authors collected background and qualitative and quantitative information of the company from its participation in the Best Performance in Sustainable Luxury in Latin America Award, getting this award in the category 'Best Fashion and Accessories Enterprise of Sustainable Luxury' in the 2011 edition. The criteria needed to be met to receive the award are:

- *social aspects*: the strategies carried out by the company underscoring positive impacts;
- *environmental aspects*: the strategies carried out by the company underscoring positive impacts;
- *economic aspects*: upfront investment; sales volume, profits (as a percentage of revenues), future growth expectations based on company performance, average price of product/s that your company sells, and distribution or sales channel

This source was supplemented with the chapter written by Kavita Parmar (2014) in the book titled Sustainable Luxury and Social Entrepreneurship - Stories from the Pioneers, whose editors were Miguel Angel Gardetti and María Eugenia Girón (2014). Moreover, the authors considered the e-mails exchanged with Ms. Parmar within the framework of the above award.

The analysis of the collected information focused on the business model based on the canvas presented by Clark et al. (2010). Even though this model not only applies to the clothing industry, it puts special emphasis on the supply chain development and on the relationship developed between the initiative and its customers. It was also analyzed if within the framework of the canvas the initiative fostered a relationship between the customer and the artisans who were part of the supply chain.

(Disruptive) innovation and business models

Innovation provides the means to create a new reality. It involves converting knowledge, learning, capabilities, and insights into value and creative new perspectives, products, productive outcomes, and business models. Innovation is more than change: it is making incremental and/or radical improvements to systems, technologies, products, processes, and practices (business models) (Rainey, 2004).

There are different types of innovation, namely sustaining, evolutionary, revolutionary, and finally disruptive (Christensen, 1997) innovation. Out of all of them, disruptive innovation is very much interesting and quite applicable to apparel industry. It is rather the need of the hour for today's apparel sector. Disruptive innovation is defined as innovation which helps creating a new market (and eventually disrupts an existing market) by implementing different value sets (Christensen, 1997; Christensen and Raynor, 2003; Christensen and Overdorf, 2004). Disruptive innovation refers not to the incremental improvement but rather to innovation that leapfrogs standard routines and knowledge (Hart and Milstein 1999). It is what Quinn (1996) called a much focused process - discontinuous with the past and generally irreversible. Thus, rather than simply seeking to reduce the negative impacts of their operations, firms can strive to solve social and environmental problems through the internal development or acquisition of new capabilities that address the sustainability challenge directly (Hart, 1997, 2005a, 2007 and Hart, 2003). It is what Rainey (2004) defined as moving from thought to action.

Disruptive technologies can bring to the marketplace a very different value proposition than previously available, and, most importantly, products based on disruptive technologies can be cheaper, simpler, and smaller and can be more frequently made and very convenient to use (Christensen, 1997; Aneja, 2010).

The business model concept has become increasingly used to provide explanations and tools for studying the dynamics of businesses (Amit and Zott, 2001; Zott et al., 2011). Several authors have proposed varying definitions and theoretical frameworks to explain business models. See for example the works of Perkmann and Spicer (2010), Osterwalder

(2004), Osterwalder et al. (2005), and Teece (2010). Osterwalder and Pigneur (2010) have applied a pragmatic perspective to the concept that helps to understand how a firm does business, for analyses, comparison, performance assessment, management, and communication, and to assist firms in their innovation. Given this provenance, some concepts prioritize the creation of economic value for business (Chesbrough and Rosenbloom, 2002).

Lüdeke-Freund (2010) describes a sustainable business model as a business model that creates a competitive advantage through superior customer value and contributes to a sustainable development of the company and society. The business model concept has recently been employed in the context of sustainable innovation (Boons and Lüdeke-Freund, 2013, Wüstenhagen and Boehnke, 2008, Wells, 2008, Hannon et al., 2013).

A frequently used approach from Osterwalder and Pigneur (2010) deconstructs the business model into nine inter-related 'building blocks.' These blocks require:

- Customer segments: customers comprise the heart of any business model. An organization must take a conscious decision about which segments to serve and which segments to ignore. Once this decision is made, a business model can be carefully designed around a strong understanding of specific customer's needs.

- Specification of the value proposition: the value proposition is the reason why customers turn to one company over another. The value proposition consists of a selected bundle of products and/or services that caters to the requirements of a specific customer segment.

- Channels: are customers' touch points that play an important role in the customer's experience.

- Customer relationships: a company should clarify the type of relationship it wants to establish with customer segment. This relationship may be driven by customer acquisition, customer retention, and boosting sales.

- Revenue streams: represents the cash a company generates from each 'customer segment'.

- Key resources: these resources allow an enterprise to create an offer a 'value proposition', reach markets, maintain relationship with 'customer segments', and earn revenues.

- Key activities: these are the most important actions a company must take to operate successfully.

- Key partnerships: companies create alliances to optimize their business models, reduce risk, or acquire resources.

- Cost structure: this is the most important cost incurred while operating under a particular business model.

Figure 1 shows the CANVAS model developed by Osterwalder and Pigneur (2010). It should be noted that since Osterwalder and Pigneur (2010) defined customers as the heart of any business model, we adjusted the original version, putting customers in the first place.

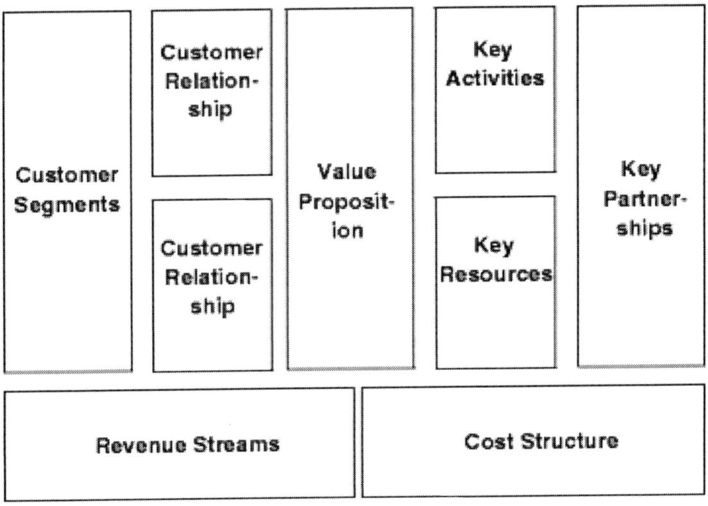

Figure 1. CANVAS model developed by Osterwalder and Pigneur (2010). Source: prepared and adapted by the authors (based on Osterwalder and Pigneur (2010)).

In 2012, Clark - with the cooperation of Osterwalder and Pigneur- adapted the concepts of the 2010 CANVAS to business models where the key resources are, basically, individuals (entrepreneurs), their skills and abilities, and the assets they own and control (Clark et al. 2012). Based on this adjustment, Figure 1would be as follows (Figure 2).

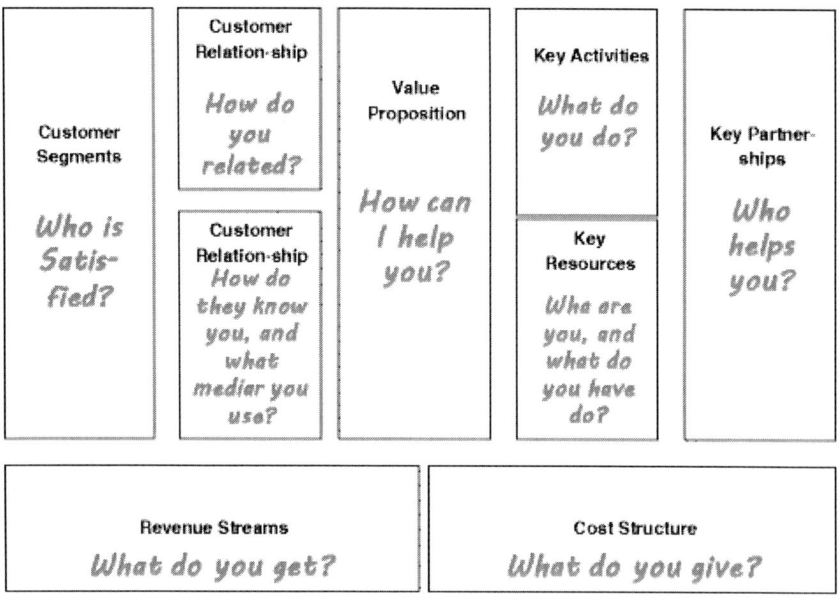

Figure 2. Personal CANVAS model developed by Clark et al . (2012). Source: prepared and adapted by the authors (based on Clark et al. (2010)).

2.1 The IOU Project[a]

Kavita Parmar and Iñigo J. Puente Henales (her husband) created the IOU Project. With an investment of USD 2.5 million - between the founders and financial funders - and profits of USD 1.5 million (in 2012)[b], IOU is a new disruptive e-commerce initiative that offers fashion-conscious consumers a very distinct platform[c]. This platform not only enhances the online experience by linking end buyers with artisan producers but also draws upon the most compelling aspects of social media to construct a community around its core brand values: authenticity, transparency, uniqueness, and both social and environmental responsibility. Born from the need to empower both the artisan and the consumer, it uses the full scope of modern technology and the social web by creating a new supply chain that we call the 'prosperity chain', and IOU expects a 100% growth in the coming years based on the past years' performance.

The IOU's mission is to promote responsible consumption, whereby the consumer engagement is fundamental. People are already very much interested in the provenance of things; once they see who makes the product they use, there is an emotional value attached to knowing its true worth. This makes the product less disposable.

IOU is wholesaling the product directly to their consumers, which means that IOU can make the highest quality product paying the craftsman in India and Europe their fair share and still sell it at the same price as most 'high street retailers' so that the consumer gets great value at an affordable price tag. A comparable shirt made in Italy with handmade fabric would cost at least twice the price at the IOU's site. Prices range from USD 30 to 170.00. Thus, consumer's purchases are allowing these weavers and artisans in India and Europe to self-sustain and not have to migrate hundreds of miles to work in factories as laborers.

But, most importantly, IOU's impact is that it gives back 'pride' to the artisan's work; to date, these artisans have never had a chance to speak, let alone get connected with end consumers. In this relationship, the consumer can learn some things from the artisans and have access to a video that introduces them as well. For example, the weaver artisan R. Ramanathan[d] belongs to a cooperative called Kanjamanathanpettai Handloom Weavers Cooperative and has been a member of IOU since 26 September 2010. The website also shows some characteristics of the cooperative to which this artisan belongs. For instance, the year it was founded, how many artisans it employs and how many looms they have, etc. In turn, the weavers who are the members of the IOU Project - who have agreed to work with higher quality standards - receive economic compensation which is equal to 100% of their usual salary for each lungi[e] they knit. An interesting testimonial that shows the possibilities this relationship entails is that of Jessica Adams, a consumer and textile design student from the University College Falmouth (England), who visited Karaikadu Handloom Weavers Cooperative. On the last day of her visit, she said: '*It was a lovely day, the set up here seems to be working well and the weavers seem to be content. The atmosphere is so wonderful. I really hope to return one day, and wish all these wonderful people the best for the future...Hopefully this weaving way of life can continue*'.

The IOWEYOU label has taken beautiful hand-woven 'Madras' plaid fabrics made for centuries by artisans in India and turned them over to artisans in Europe who create one-of-a-kind apparel items to be sold online via the IOU specialized social e-commerce tools, the 'trunk show host.' The affordable, easy-to-wear apparel items found on the site are handmade and unique, in every sense of the word.

On the IOU website, buyers can shop for unique items and have access to a wealth of audiovisual material that chronicles the process whereby an item was produced. End buyers also have the opportunity to 'become part of the story' of each item by using a unique QR code on the items which electronically links them to the specific artisans that created their unique piece. IOU has already expanded into other authentic fabric sources and artisan groups: the real indigo plant-dyed denim from Japan, real cashmere wool from Northern India/Nepal, and yak wool from Tibet, among others. All of them sourced from artisan communities with full traceability and transparency. The IOU big dream is to create a Wikipedia of artisans worldwide.

The IOU Project also offers partner retailers, and others brand the possibility of using the platform and the IOU's MAAP (mass aggregated artisan produced) system to produce their designs.

3. RESULTS AND DISCUSSION

3.1 The IOU Project analysis through Clark, Osterwalder, and Pigneur's personal CANVAS (2012)

Below, we will analyze the case of the IOU Project based on Clark et al. (2012) CANVAS structure:

Who is satisfied?

IOU knew that there was a small part of the audience that cared about the ideas of traceability and transparency, but it is intended to reach out a larger audience. In Kavita's word: '*I was obsessed with the idea of addressing a larger market, that you could buy a piece at a reasonable price point and that it was truly unique felt like important sales arguments.*' However, in order to increase its market share and promote craftsmanship, the IOU Project might consider the ideas stated by Lynn Stephen in her paper 'Export Markets and their Effects on Indigenous Craft Production: the Case of the Weavers of Teotitlán del Valle, México' (1996:382). The international market for crafts is built on a consumer ideology that contrasts manufactured, mass-produced, modern objects with handmade, authentic, and local crafts, which means that these artisans and traders do not necessarily have to adopt the norms of capitalist relations to operate in these markets (Lockwood 1993).

Here, the purchase goes beyond the object itself; excellence and devotion give both life and a sense of deep connection between individuals.

How do you relate?

According to the founder - Kavita Parmar: '*Each IOWEYOU Madras piece is unique and has an IOU code(QR Code) that takes you to its story simply by scanning it with a Smartphone or just entering the number on our website. On the site you can see the weaver who hand wove that particular fabric in India and then the artisan in Europe who turned that fabric into a beautifully handcrafted piece. We encourage the customer to take their picture and upload it to complete the story so that the weaver in India and the artisan in Europe can also see who bought that piece.'* The founder of the IOU Project encourages consumers to 'become part of the story.' This is possible because consumers are not ignoring the price/quality ratio but are applying and prioritizing other criteria in their purchase decision process (Harrison et al., 2010:2).

How do they know you and what media do you use?

Iñigo J. Puente Henales - Kavitas' husband - is an engineer from MIT and has always had the belief that with the current technology, IOU could have the tools to truly disrupt the fashion industry. Kavita explains it like this, '*the consumer does care and the proof is in the pudding. Being a self-funded start up and this being a new idea that took a lot of time and resources to develop, when we went live we had no money left for marketing or PR. But people spread the word like wild fire. We continue to receive emails from people around the globe who love the idea and want to support us. Also we added the very consumer engaging trunk show host feature. Since as a company we were being disruptive with the current supply chain,why not take that one step further and do the same experiment with distribution of the brand. On our website people who are truly interested in the project can have their own trunk show.'* This evidences that the IOU Project moved from the traditional 'brand-centric' concept to focus on a consumer wanting more direct communication and dialogue with the company. According to Rinaldi and Testa (2014), this revolution was inevitable given social networks and increasing transparency, thanks to the online presence of company information which might be of interest to the end consumer.

How can I help you?

A 'story' as Kavita said, '*I wanted the first collection to be around an authentic story and what could be more authentic than Gandhi's non-violent revolution? He inspired millions of Indians to buy local handmade fabrics to fight against the cheaper machine-made imports from imperialistic 19 th century England to give back work to the local craftsman.'* Sustainability is redefining the world of textiles and fashion. Therefore, the IOU Project offers a 'story' instead of a garment. That was precisely one of the seven demands addressed by young (undergraduate and postgraduate) students (from all over the world) during the Youth Fashion Summit, which were in turn submitted to the industry at the Copenhagen Fashion Summit, 2014.

What do you do?

Kavita explains it like this, '*In the last 20 years the focus of all factories and ateliers has been to increase productivity, as they had to compete with global sourcing. We had to go back to how things were made before when a very small, reduced group made the entire piece. This gave us better quality and traceability. What was magical was that it also gave back authorship and brought back pride to the artisan. Creating the online and offline traceability system was one of the most challenging tasks but it has also been the most rewarding. Making people aware of each other in the chain has been one of the true successes of the project, something that I cannot describe on a balance sheet but have witnessed repeatedly in the proud smiles of the artisans when they see the Prosperity Chain on the web. Our focus is on value and quality. There is a huge imbalance in people's perception of value. The industry has convinced the consumer to accept low quality for a cheaper price point and turned everything into a disposable product*'.

By creating this *prosperity chain*, the IOU Project is sustaining artisans' deeply held beliefs about their social relations and their relationships to the environment (Grimes and Milgram, 2000: 7). This is so because their legitimacy as craft items hinges on the circumstances of their production as they are made using traditional craft techniques (Stephen 1996:393).

Who are you and what do you have?

Kavita Parmar is a self-taught designer and a serial entrepreneur. The IOU Project was born out of Kavita Parmar's frustration as a designer with the current fashion system. It is all about *faster* and *cheaper*; it has become a race to the bottom and does not nurture *big design* (not just making things pretty but truly designing the ecosystem around their production to make it sustainable) nor value provenance, excellent craftsmanship, or artisanship.

Fletcher and Grose (2012) analyzed the roles that should be played by designers within the sustainability framework, among which we find that of communicator, facilitator, and activist. Throughout the development of this project, Kavita Parmar is playing all of these roles. Moreover, she regard this as a 'cultural' project that gives a sustainable answer to artisans' employment and contributes to artistic preservation (Liebl,2005).

Who helps you?

Focusing on what was most important to Kavita, the artisan who helped to make IOU designs a reality and the customer who bought them. So IOU worked on the first collection with handloom weavers from Tamil Nadu, which were founded in 1927 inspired by Mahatma Gandhi. But, in addition to this, IOU wanted to produce the clothing with the best makers and Europe with its artisanal tradition.

For the IOU Project, craftsmanship is an essential principle (while for its 'fans', it is an essential driver for their purchase decision). Craftsmanship is an essential part of the IOU Project's rationale. It is a value-creating differentiator. That is the reason why IOU features its products as 'unique' (each garment is presented as a one-of-a-kind creation). It is the world of unique tailor-made products and limited editions. This concept brings customers closer to artisans.

What do you get and what do you give?

Using transparency and traceability, the IOU Project aims to turn the supply chain into what we call a*prosperity chain*: IOU buys from weavers and artisans at the fair price they ask and works on low mark-ups compared to the normal industry standards to ensure that the final price is competitive. Table 1 includes a summary of the main aspects of the IOU Project.

Table 1. The main aspects of the IOU Project

	The IOU Project
Size (sales)	$ 1,500,000.00
Profit	Break even 2013/2014
Expect growth based on previous years	100%
Average price	$ 30 - 170
Distribution	B2C
Investment	$ 2,500,000.00
Investors	Founders + FF
Social impact	Building prosperity chains, where products are embedded with full traceability from artisan to consumer. Pride and higher prices for craftspeople.
Environmental impact	Locally grown cotton

Source: prepared by the authors.

4. CONCLUSIONS

Most organizations - whether start-ups or established companies - take very defensive postures that protect and nurture all that which has served them well in the past, even in the face of change (Teece et al. 1997; Christensen, 1997). That is to say, *they react by increasing their commitment to the 'existing' products,processes, and markets*. However - according to Hart and Milstein (1999) - successful surviving companies or start-ups have been able to invest or partner in order to achieve new competencies and experience in new untapped markets. That is the difference between continuity and discontinuity as pointed out by Hart (2005b). The former is illustrated by concepts such as: efficiency, operations, control, resource allocation, and convergent thinking. The latter is illustrated by: imagination, innovation, creative destruction, resource attraction, divergent thinking. It is also what Milstein, Hart, and London (2007) called the 'revolutionary routines.' It is what Kavita Parmar understood by 'breaking' with the dominant system in the textile and fashion industries.

As a business model describes the rationale of how an organization creates, delivers, and captures value (Osterwalder and Pigneur, 2010: 14), a transformation of dominant business models is, therefore, needed in order to achieve systemic changes toward sustainability. Rather than perceiving sustainability solely as compliance and risk management, businesses have to recognize the value-creating potential of sustainable business models. Using a disruptive business model, the IOU Project manages to develop a value chain based on craftsmanship. The dawn of this shift makes individual *savoir-faire* more important and valuable than ever. Craftsmanship is about women and men imagining and creating, failing and succeeding, searching and finding, and preserving and challenging - and developing unique expertise in the process. It is, just as importantly, about the constant flow of such expertise from one generation to the next, across decades and centuries (Ricca and Robins, 2012: 53). Craftsmanship breathes humanity into objects, allowing the individual to create for the individual. In doing so, it translates the principles of knowledge, purpose, and timelessness into products that represent unique achievements (Ricca and Robins, 2012: 53). Within this framework, supply chain management became a strong driver of value, for the IOU but for artisans and customers as well (Kell, 2010). The IOU Project's approach includes associating with artisans to innovate and provide for sustainable win-win scenarios (Hammond, 2001; Prahalad and Hammond, 2002; Prahalad and Hart, 2002; Hammond 2004a, b; Hart, 2005a; Prahalad, 2005). Limitation of the analysis presented in this paper with the aid of IOU Project, this analysis is based on the perspective of the business owners without much reference to the artisans' perspective of how it works for them.

The IOU Project's business model helps to develop a strong relationship with consumers and artisans which dignifies the later and appreciates their work. Figure 3 shows these relationships within the framework of Clark et al. (2012) CANVAS: the blue arrow with number 1 shows the relationship that the IOU Project developed with its clients. The blue arrow with number 2 shows the relationship with consumers that share the same values that the IOU Project fosters. The light blue arrows with number 3 show the close relationship with artisans. And the arrows with number 4 show the relationship built between consumers and artisans through the IOU Project website.

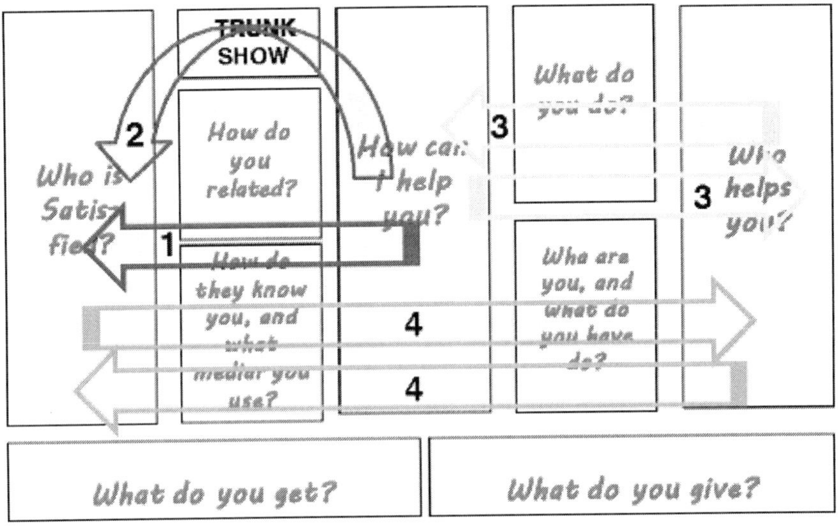

Figure 3. Relationships established by the IOU Project through the personal CANVAS model developed by Clark et al. (2012). Source: prepared and adapted by the authors (based on Clark et al. (2010)).

ENDNOTES

[a]This portion is based on Gardetti and Girón (2014) Sustainable Luxury and Social Entrepreneurship - Stories from the Pioneers; and information during the process of the second edition of 'Best Performance in Sustainable Luxury in Latin America Award' (at present *IE Award on Sustainability in the Luxury and Premium Sectors* at global level); and on exchange of e-mails with Mrs. Kavita Parmar - the IOU Project founder.

[b]IOU Project break even will be 2013/2014.

[c]IOU website: http://iouproject.com/

[d]The IOU Project (2015): http://iouproject.com/stories/weaver/ramanathan/ Accessed 5 Jan 2015.

[e]Each garment is made from a different woven piece, which is 2 m long, called lungi, also known as 'Madras' square. The lungi is 2 m long and has been woven in India by artisans for centuries.

COMPETING INTERESTS

The authors declare that they have no competing interests.

AUTHORS' CONTRIBUTIONS

The authors read and approved the final manuscript.

REFERENCES

1. Amit, R. and C. Zott (2001) Value creation in e-business. Strategic Management Journal, 22, 493–520.
2. Aneja, A, 2010., The art (and science) of propelling growth in textiles, 7th International Conference - TEXSCI 2010, September 6–8, Liberec, Czech Republic, Available at: http://wjoe.hebeu.edu.cn/sup.2010/Fibre%20Science%20and%20Smart%20Textiles/9_o_Aneja.pdf
3. Boons, F, & Lüdeke-Freund, F. (2013). Business models for sustainable innovation: state-of-the-art and steps towards a research agenda. *Journal of Cleaner Production, 45*, 9–19.
4. Clark, T, Osterwalder, A, & Pigneur, Y. (2012). *Business model you.* New Jersey: John Wiley & Sons.

5. Christensen, CM. (1997). *The innovator's dilemma: when new technologies cause great firms to fail.* Boston: Harvard Business Press.

6. Christensen, CM, & Raynor, ME. (2003). *The innovator's solution: creating and sustaining successful growth.* Boston: Harvard Business Press.

7. Christensen, CM, & Overdorf, M. (2004). Meeting the challenge of disruptive change. In RA Burgelman, CM Christensen, & SC Wheelwright (Eds.), *Strategic management of technology and innovation - fourth edition*(pp. 541–549). New York: McGraw-Hill Irwin.

8. Chesbrough, HW, & Rosenbloom, RS. (2002). The role of the business model in capturing value from innovation: evidence from Xerox Corporation's technology spinoff companies. *Industrial and Corporate Change, 11*, 533–534.

9. Fletcher, K, & Grose, L. (2012). *Fashion & sustainability: design for change.* London: Laurence King Publishing.

10. Gardetti, M.A. and M.E. Girón (eds) Sustainable Luxury and Social Entrepreneurs: stories from de pioneers. Greenleaf Publishing, Sheffield.

11. Grimes, KM, & Milgram, DL. (2000). Introduction. In KM Grimes & B Lynne Milgram (Eds.), *Artisans and cooperatives: developing alternate trade for the global economy* (pp. 3–10). Tucson: The University of Arizona Press.

12. Hammond, A. (2001). Digitally empowered development. *Cooperation South Journal, 1*(1), 5–15.

13. Hammond, A. (2004a). Unleashing entrepreneurship among the poor. *Sustainable Development International, 12*(1), 1–3.

14. Hammond, A. (2004b). *Technology, globalization and the poor*. Washington: World Resources Institute, Washington.

15. Hannon, MJ, Foxon, TJ, & Gale, WF. (2013). The co-evolutionary relationship between energy service companies and the UK energy system: implications for a low-carbon transition. *Energy Policy, 61*, 1031–1045.

16. Harrison, R, Newholm, T, & Deirdre, S. (2010). Introduction. In R Harrison, T Newholm, & S Deirdre (Eds.), *The ethical consumer* (pp. 1–8). London: Sage.

17. Hart, SL. (1997). Beyond greening - strategies for a sustainable world. *Harvard Business Review, 75*(1), 66–76.

18. Hart, SL. (2005a). *Capitalism at the crossroads*. Upper Slade River: Wharton School Publishing.

19. Hart SL (2005b) Creative destruction y sustainability, lecture delivered within the framework of the Argentina Base of the Pyramid Learning Lab (Center for Study of Corporate Sustainability, Argentina, September 2005)

20. Hart SL (2007) Capitalism at the crossroads - capitalism at the crossroads - aligning business, earth, and humanity. 2nd ed. Wharton School Publishing, Upper Slade River

21. Hart, SL, & Milstein, M. (1999). Global sustainability and the creative destruction of industries. *MIT Sloan Management Review, 41*(1), 23–33.

22. Hart, SL, & Milstein, M. (2003). Creating sustainable value. *Academy of Management Executive, 17*(2), 56–67.

23. Kell, G. (2010). Prologo. In S Cody Sisco, C Bythle, & P-J Peder Michael (Eds.), *Supply chain sustainability: a practical guide for continuous improvement* (p. 2). New York: United Nations Global Compact.

24. Liebl M (2005) Jodhpur: The most creative of communities. Paper presented at the Conference on Asia-Pacific Creative Communities: A Strategy for the 21st Century Senior Expert Symposium, Jodhpur February 22–26, 2005.

25. Lockwood, VS. (1993). *Thaitian transformations: gender and capitalist development in a rural society*. Boulder: Lynne Rienner.

26. Lüdeke-Freund F (2010) Towards a conceptual framework of business models for sustainability. ERSCP-EMSU Conference 2010 - Knowledge Collaboration and Learning for Sustainable Innovation. Centre for Sustainability Management (CSM), Leuphana University of Lüneburg.

27. Martin M (2013) Impact Economy, 2013, Creating sustainable apparel value chains: a primer on industry transformation. Suisse: Impact Economu. Available at: http://www.impacteconomy.com/papers/IE_PRIMER_DECEMBER2013_EN.pdf

28. Milstein, M, Hart, SL, & London, T. (2007). Revolutionary routines. In SK Piderit, RE Fry, & DL Cooperride (Eds.),*Handbook of transformative cooperation - new designs and dynamics*. Stanford: Stanford University Press.

29. Muthu, SS. (2014). *Assessing the environmental impacts of textiles and the clothing supply chain*. UK: Woodhead Publishing.

30. Osterwalder A (2004) The business model ontology: a proposition in a design science approach. Institut d'Informatique et Organisation. Lausanne, Switzerland, University of Lausanne, Ecole des Hautes Etudes Commerciales HEC

31. Osterwalder, A, & Pigneur, Y. (2010). *Business model generation: a handbook for visionaries, game changers, and challengers*. New Jersey: John Wiley.

32. Osterwalder, A, Pigneur, Y, & Tucci, CL. (2005). Clarifying business models: origins, present, and future of the concept.

Communications of the Association for Information Systems, 16(1), 1–25.

33. Parmar, K. (2014). The IOU Project. In MA Gardetti & ME Girón (Eds.), *Sustainable luxury and social entrepreneurship - stories from the pioneers*. Sheffield: Greenleaf Publishing.

34. Perkmann, M, & Spicer, A. (2010). What are business models? Developing a theory of performative representations. In: Nelson Phillips, Graham Sewell, Dorothy Griffiths (eds.) Technology and Organization: Essays in Honour of Joan Woodward. *Research in the Sociology of Organizations, 29*, 265–275.

35. Prahalad CK (2005) The fortune at the bottom of the pyramid. Wharton School Publishing Upper Slade River

36. Prahalad, CK, & Hart, SL. (2002). The fortune at the bottom of the pyramid. *Strategy + Business, 26*(2002), 54–67.

37. Prahalad, CK, & Hammond, A. (2002). Serving the world's poor, profitably. *Harvard Business Review, 80*(9), 48–57.

38. Quinn, R. (1996). *Deep change: discovering the leader within*. San Francisco: Jossey-Bass.

39. Rainey DL (2004) Sustainable development and enterprise management: creating value through business integration, innovation and leadership. Article presented at Oxford University at its colloquium on Regulating Sustainable Development: Adapting to Globalization in the 21st Century - August 8 through 13, 2004.

40. Ricca, M, & Robins, R. (2012). *Meta-luxury: brands and the culture of excellence*. London: Palgrave MacMillan.

41. Rinaldi, FR, & Testa, S. (2014). *The responsible fashion company: integrating ethics and aesthetics in the value chain*. Sheffield: Greenleaf Publishing.

42. Schaltegger S, Ludeke-Freund F, Hansen, EG (2011) Business cases for sustainability and the role of business model innovation: developing a conceptual framework. Centre for Sustainability Management, Leuphana University of Lueneburg, Germany

43. Sneddon, C. & Howarth, R.B. (2006) Sustainable development in a post-Brundtland world, Ecological Economist, no. 57, pp. 253–268. Available at: http://www.sciencedirect.com/science/article/pii/S0921800905002053.

44. Stephen, L. (1996). Export markets and their effects on indigenous craft production: the case of the weavers of Teotitlán del Valle, México. In S Margot Blum, JC Berlo, & EB Dwyer (Eds.), *Textile traditions of Mesoamerica and the Andes: an anthology* (pp. 381–402). Austin: University of Texas Press.

45. Stubbs, W, & Cocklin, C. (2008). Conceptualizing a "Sustainability Business Model". *Organization & Environment, 21*(2), 103–127.

46. Teece, DJ. (2010). Business models, business strategy and innovation. *Long range planning, 43*(2), 172–194.

47. Teece, DJ, Pisano, G, & Shuen, A. (1997). Dynamic capabilities and strategic management. *Strategic Management Journal, 18*(7), 509–533.

48. World Commission on Environment and Development. (1987). *Our common future*. Oxford: Oxford University Press.

49. World Economic Forum. (2010). *Redesigning business value: a roadmap for sustainable consumption*. Geneva: World Economic Forum (WEF).

50. Wells, P. (2008). Alternative business models for a sustainable automotive industry. In A Tukker, M Charter, C Vezzoli, E Stø, & M Andersen (Eds.), *System innovation for sustainability*. Sheffield: Greenleaf Publishing.

51. Wüstenhagen, R, & Boehnke, J. (2008). Business models for sustainable energy. In A Tukker, M Charter, C Vezzoli, E Stø, & M Andersen (Eds.), *System innovation for sustainability*. Sheffield: Greenleaf Publishing.

52. Zott, C, Amit, R, & Massa, L. (2011). The business model: recent developments and future research. *Journal of Management, 37,* 1019–1042.

2

THE TRUE VALUE OF MATERIALS: BRIDGE (BUILDING RESEARCH AND INNOVATION DEALS FOR THE GREEN ECONOMY)

Joan M. Farrer* and Carolyn A. Watt

Faculty of Creative and Cultural Industries, University of Portsmouth, Portsmouth, PO1 2DJ, UK

ABSTRACT

Assessing the value of materials, lifecycle and applications was central to the European Union INTERREG IV project BRIDGE (Building Research and Innovation Deals for the Green Economy) 2011–2014. Here, the complex philosophies of sustainability (protection of people, profit and planet) underpinned innovation, knowledge transfer, data visualization and design thinking, to develop green entrepreneurs, and market differentiation in Southern England and Northern France Channel regions.

The model followed transdisciplinary collaborative research methods resulting in 'green best practice', which was visualized and disseminated extensively via digital technologies, mixed media and exhibitions. Information generated by the research was debated and shared at cross-channel conferences, business-to-business, incubation events, workshops and research exchange visits focussing on two materials: textiles and wood, sourced in the UK and France. Partners came from universities, regional authorities, non-governmental organisations and business communities to discuss the economic, environmental and societal value of these regional materials and innovative proofs of concept, eco and sustainable design products, processes, services and material experiments were generated from this collaborative peer learning, community of practice approach.

More sustainable proof of concept products and systems were developed by knowledgeable practitioners which embody an understanding of green business, where the philosophical rationales and complexity of ethics, climate change and waste issues, for example, were communicated through materials and objects to expert and non-expert audiences and consumers. This embodiment and honest communication of knowledge in a product is a unique selling point, creating market value and consumer differentiation through narrative. Stakeholders exchanged complex data, methods and ideas towards developing green employment opportunities, informing a research theme for UK and EU funders to the year 2020 and beyond.

This is a descriptive narrative on the evolvement of the project which enables incisive, reflective and theoretical analysis to take place simultaneously in other publishing areas to assess the longer term impact and value of BRIDGE in subsequent transdisciplinary projects. This work contributes to literature on value and use of materials with a focus on collaboration, design, innovation, applied research and societal benefit to develop green employment.

KEYWORDS

Design Innovation, Materials Knowledge transfer, Green Sustainable Transdisciplinarity, Textiles Business

1. BACKGROUND

This paper reflects upon the aims, objectives and outcomes of the 2.4 million euro-funded project BRIDGE (Building Research and Innovation Deals for the Green Economy)1 which was informed by Brundtland's classical description of sustainability:

'Development that meets the needs of the present without compromising the ability of future generations to meet their own needs' 2

The intentions for this article take a very particular form: in that it describes aspects of what took place in the complex BRIDGE project and provides the (vital) contextual platform for further discussion and aligns with other project publications with different foci taking place simultaneously to produce a cluster of outputs. In the field of creative practice there is a tradition of describing the narrative of a project's evolution—which is the aim of this article and separates descriptions of research process from analysis and theorizing. However, the scale and scope of the project makes it difficult to steer a path through this description, where reflection and analysis in such a recently concluded project is not its purpose.

The question was how to develop a new generation of creative and business thinkers, thinkers that would be knowledgeable considering the sustainable (or green) agenda; a global, multifaceted and complex subject. For the purpose of understanding the larger concept of sustainability, the BRIDGE project focussed upon two local material streams common in France and the UK where material lifecycles were analysed via the cradle-to-cradle concept (Braungart and McDonough 2009). The analysis of the lifecycle of two materials functioned as a scalable knowledge toolbox to interrogate sustainable philosophies and processes and practice to up-skill stakeholders. Timber and textile material production, products and disposal became the basis of the study where artefacts to visualize data became a significant part of project deliverables during January 2012 to December 2014. The project partnership consisted of design, science, technology, engineering, mathematics and business colleagues (the D-STEM-B model, conceived by lead author Farrer in 2011). The intention of the process was to develop a mixed methods skill toolbox from the partners' best practice, where outcomes from each discipline could be shared, discussed, considered and applied to products and systems for existing and new markets. The analysis and application of the concepts, both practical and theoretical, from the mixed methods toolbox was estimated by the project leaders to contribute to a deeper understanding and up-skilling in sustainable production and consumption issues. All those involved in this research project were informed by relevant *eco supply chain narratives* the knowledge about which was subsequently transferred to the thousands of people, who engaged with the work (evidenced in feedback) at BRIDGE exhibitions, conferences, seminars, business-to-business (B2B) events, media exposure and through social networks.

The lead partner was a French government development agency MIRIADE (Mission Régionale pour l'Innovation et l'Action de Développement Économique, France), who devised the overarching BRIDGE concept for green development to include SMEs (small and medium enterprises), HEIs (higher education institutions) and NGOs (non-governmental organisations) as a French and English collaboration to secure EU funding. The partner's expertise was in the material fields of research, agriculture, product design, retail, and disposal in the UK and French regions and those who joined the consortium outlined their contribution to and aims for the project. Finally, a core of nine partners across regions and fields of expertise was consolidated through workshop debates and email exchanges to develop the concept of Building Research and Innovation Deals for the Green Economy, BRIDGE. The partnership breadth enabled knowledge transfer amongst the group, including business incubation support and enabling the practical development of more sustainable materials products and services to inform the creation of sole traders, SMEs and non-profit organisations for an increasingly aware European green consumer. The effects of linking these partnerships, facilitating new knowledge and access to eco markets could increase employment—which was at the heart of the project. The consortium used research literature, symposia, exhibitions, field work, workshops, B2B events and practice to discuss and investigate lifecycle analysis, improve ecological material processes, and zero waste, re-use, up-cycling and closed loop practices. Sustainable design thinking was one of the core components that needed to be understood and adopted as fully as possible by the partners, green businesses and education. A complex digital communication toolbox for partners, commercial contacts and the public was established to map producers in the geographical locations and explain to interested parties the projects' rationale and development. Web links to information streams and social media increased engagement and knowledge transfer between the communities of learners who were at various stages of their understanding.

The INTERREG funders (Joint Technical Secretariat, European Commission) monitored the deliverables for the BRIDGE project work packages every 3 months over the 3-year period. The deliverables included eco-design research, development and innovation; exhibitions and workshops; new research and development projects and product outcomes; hosting cross-channel gateways and symposia; forming a database and live website for green growers, makers, users and entrepreneurs in southern England and northern France; a mapping of primary and secondary material resources and student exchanges and bilateral agreements between French and UK universities, 10 new businesses and research and development outcomes.

2. CASE DESCRIPTION

2.1 Operating strategy: an account of the BRIDGE project—mechanisms and methods

The BRIDGE project follows a transdisciplinary approach, which is suitable for a project interested in environmental, economic and social issues because it is a method that recognizes wholeness and the value of otherness. It was philosopher Jean Piaget who first presented the theory of transdisciplinarity in a lecture, in 1977, which described a connection between and beyond disciplines, which worked together to provide transparent sets of methods of communicating work and data to each other outside of their respective discipline silos. Describing complex concepts such as sustainability needs a web or network of information, with many different strands: Piaget found one phrase that could make all the connections: transdisciplinarity. This multimethod approach rather than one meta-method was critical for BRIDGE collaborators, who brought a range of expert research and knowledge to the project through critical debate, exploring material technology and creative practice leading to new understanding and innovation expressed by Mills (2014):

'As technology progresses, there is the need for increasing specialisation. There is also the need for increased collaboration amongst those with specific expertise. This is one of the reasons that networks (in their broadest sense) assist with innovation—and their importance continues to grow.'

This statement defends the use of a complex partner network such as this in the BRIDGE project where the intention was to provide a unique development opportunity for new cross-border collaborative methods and opportunities to cascade best practice, knowledge and experience in ecological terms. The knowledge gleaned from conducting the project could be applied across geographical areas to support those committed to developing green businesses and new markets in UK and French regions with high unemployment due to industrial and agricultural decline. Using materials as the focus for dialogue and innovation, local authorities, research, eco design, business incubation, production, biotechnology, agriculture and waste reprocessing could come together to debate and facilitate opportunities for research development and collaboration. In addition, communicating the complexity of sustainability to a non-expert audience in countries through visual 3D and virtual artefacts, formal and informal events and Internet presence was a challenge. Communication of the aims and objectives and outcomes is deemed to be a success following analysis of qualitative unstructured interviews, written and verbal commentary and anecdotal evidence from the public, learners and professionals who engaged with the project. The impact from these knowledge exchange and sharing events enabled transdisciplinary thinking, resulting in information up-skilling for sustainable design solutions. The products created market differentiation and built consumer confidence through an honest narrative on material origin, lifecycle and product transparency, important for customers sceptical of the green credentials and 'green wash' of some brands. It was noted that the importance of a transparent, environmentally and ethically sound business or developing a sustainable mission statement and philosophy for a brand was not properly understood in each country or by each partner. For instance, the French partners did not have an equivalent for the word or meaning of sustainability and used a combination of *green, eco, environmental* or *ecological* terms, where the focus was in general on quantitative measurements. From a UK partner perspective, sustainability combined societal, environmental and economic elements, which was informed by a combination of quantitative and less measurable qualitative knowledge. These complex convoluted holistic concepts were shared between partners in a variety of ways and caused tension in debates. This was unsurprising and a consideration of the practical epistemological implications of transdisciplinarity working. When there was a clash of philosophies and methods, it was because different discipline stakeholders come from different positions in the sustainable debate and had to explain their start point. Also, we defined our methodology as*transdisciplinarity*, which implied some type of social constructivism and so the project's sets of methods and stakeholder views were accommodated within the overall transdisciplinary framework by the concept focus on sustainable materials, which was essential to capture all debates and consolidate the disparate group. So this project became highly relevant as a cultural and social up-skilling tool for business and education in both countries and a learning curve in research method terms.

The partners for green entrepreneurial development and innovation in materials were represented by commerce, education and government from both geographical regions and were from engineering, academic institutions ESITC CAEN (École Supérieure d'Ingénieurs des Travaux de la Construction de Caen, France), Esitpa (France); from business incubation, WSX (West Sussex Enterprise, UK), Normandie Incubation (France) and Orne Development (France); from waste reprocessing, Remade Southeast (UK) and government departments Medway Council (UK, local authorities) and, as mentioned previously, MIRIADE. The lead design and material partner was the University of Brighton (UK, UoB). The core BRIDGE research team members from University of Brighton were recruited from the College of Arts and Humanities (CAH), Pharmacy and Bio-molecular Sciences (PABS) and the School of Environment and Technology (SET), creating a large interdisciplinary team of 30 design, science and business experts in their respective fields. The Principal Investigator (then director of the Design Research Initiatives DR-i) was Professor Joan Farrer, who through collaborative partner discussion initiated the concept for the UK part of the work packages from the point of view of sustainable materials, design research and innovation. Farrer built upon her research projects and international networks in sustainable and smart materials which supported the importance of sustainability and job creation which remains a major driver in Horizon 2020 EU funding policy. The BRIDGE project was people focussed, in line with the European Commission's social inclusion policies:

'Social investment is about investing in people. It means policies designed to strengthen people's skills and capacities and support them to participate in employment and social life.'

The challenge was understanding how the diverse working partnership could optimise and transfer mutual knowledge and skills to inform disciplinary and commercial practice and ultimately inform government employment and environmental policymakers through the INTERREG programme and JTS network reporting. Also defining our terms of reference and importance (e.g. eco/green/sustainable) in a bilingual context, in different cultures and societies, with a different degree of understanding was complicated as were logistics in terms of geographical distances and effective communication. The solution was a series of frequent cross-channel virtual and face-to-face exchanges facilitated by regular structured and co-ordinated meetings and visits to partners arranged around key events or gateways. These meetings enabled partners to really understand each other and were conducted in Southampton, Medway and Brighton in the UK, and Rouen, Alençon and Caen in France and were mapped out and agreed as essential in the milestone project calendar through partner discussions. These exchanges provided opportunities to get to know and understand the partners, trade formal and informal sustainable material knowledge and instigate debate. Given the wide scope and potential presented by this project and its contributors, where it was acknowledged the importance and impact could be long after the project ended, the overall approach was open-ended, experimental and iterative throughout. From their respective backgrounds (academic, scientific, engineering, design and trade) the partners brought their unique skill sets, perspectives and creative and practical working methods to the proposition of creating green commerce to develop future entrepreneurs. Various methodologies were drawn upon to target the most productive forms of communication and collaboration. These included narrative studies—not in a traditional sense; however, more as a method of gaining background information from the various partners regarding their current understanding of supply and disposal chain issues, for example. The studies of each partner then became a basis for continued improvement either individually or collaboratively through the project development, using their strengths to establish ethical purchasing policy through to using biomaterial replacements, for instance. Quantitative analysis took place whenever figures and data gathering were used within research and development of BRIDGE. Human interaction occurred through head counts and footfall at meetings, conferences, in and outside exhibitions and in institutions. Virtual interaction data gathering was done through media, social networking and broadcasting. Qualitative methods were used to understand the importance of and select case studies, including participatory and hermeneutic research. A mixed methods approach following on from the concept discussed earlier of transdisciplinarity enabled an understanding of how a diverse network project like BRIDGE, impacted upon the appreciation of the true value of materials, including waste which is also a raw material, not just in economic terms but also philosophically. Enabling collaborators to build bridges between academia and industry in the UK and France, to debate sustainable practice, deepen knowledge and share information, this led directly to a new environment for innovation and philosophical thinking. The RCUK Strategic Vision document (2011–2015) outlines the importance of developing:

'*innovations to ensure that the UK has a productive economy, healthy society and can contribute to a sustainable world.*' BRIDGE addresses this by closing '*the gap between researchers and users, with each influencing the other.*'

2.2 Study developing green material entrepreneurs: Pedagogy BRIDGE Club

The Universities in France included the following: ESITC Caen, a heavy engineering construction and architectural institution with a very traditional built environment technical remit; ESITPA, an agricultural institution, forestry, farming and bio-science; in the UK, the University of Brighton which offers all subjects was involved. Through various student teaching and learning techniques in both countries, such as lectures, workshops, debates, field trips, different discipline working collaborations and international exchanges, the BRIDGE project was used to explore and develop with students, staff and alumni the complex subject matter of sustainable design and production and consumption which is becoming an educational requisite. The students were given theoretical and practical insight to supply and dispose chain issues, ethics, material sources, production and disposal. Current sustainability literature was debated and expert speakers on the subject were interrogated which led to the community of learners contextualising their own work within recognized scholarly texts and expert practice. Knowledge of carbon transport miles, understanding water table pollution or ethical textiles production, for example, was then applied to their product design and development. An improved understanding of better material selection for example and environmentally sound methods of production and the incorporation of design for deconstruction and re-use ensued. This applied knowledge was drawn from a transdisciplinary network and formed the basis of the development of BRIDGE's green entrepreneurs working for a more informed society.

To facilitate rapid student learning and up-skilling in relation to research and innovation for the green economy at the University of Brighton, a BRIDGE Club was established, which met every 2 weeks. This was an open forum for transdisciplinary discussions, knowledge exchange, collaboration and project development and delivery. Parties from the academic and local community in fashion and textile design, design and craft, graphics, architecture and science were involved as illustrated in Fig. 1, 'The interwoven relationships arising from the BRIDGE Club'. The principal focus of these sessions was to investigate sustainable issues from primary material streams of textiles and wood.

This was innovative insofar as, through a non-elite forum, the club provided a perfect opportunity to foster challenging exchanges across and between disciplines and, in turn, highlighted the importance of interdisciplinary thinking for current and future environmental solutions. The club facilitated opportunities for participants to discuss a wide range of topics and included speakers from members of allied groups such as the Sustainability Action Network (a grassroots collective of students and staff), C-Change (a carbon reducing campaign run by staff and students), the Waste House (development of a living laboratory for ecological architectural design) and the Green Growth Platform (a network championing sustainability and supporting business growth) enabling a more unified awareness of activity within the sector that was shown across the partnership. Alumni contributed to sessions discussing positive and negative aspects of the evolution of their ecological careers and different pathways to success taken after study, which provided mentoring and knowledge exchange. The BRIDGE funding provided new materials, machinery and equipment for numerous innovative research and development proposals, experiments and techniques through workshops and knowledge transfer sessions, events and exhibitions. A series of supporting comments from students confirmed that the BRIDGE project was informing the learning community about the complexities of sustainability and altering the perception of the value of materials.

The following quotes are taken from unstructured interviews and written evidence relating to the significance of the project on these young green entrepreneurs at UoB:

'The BRIDGE Club has influenced my work in a number of ways. It has enabled me to broaden my mind in terms of designing ecologically.'

-Alice Liptrot, Textile Design Undergraduate student

'You would get different perspectives, and you would all be looking at the same materials and objects, but everyone would bring their own angle of perception towards it, which is really helpful from a creative point of view, if you are all talking about the same thing and seeing things differently.'

-Lawrence Lawson, Design and Craft Undergraduate student

Figure 1. The interwoven relationships arising from the BRIDGE Club (©July 2014, with kind permission of Marney Walker, University of Brighton, UK)

'We realised, that we have done all these crazy experiments and worked out practically how we can apply the material to different things, but you could put them all together and it could look like one big mess. The breakthrough was when we took a step back, and worked through a concept and brought in different interesting elements of what we could apply structurally, rather than putting practical techniques together to form a structure. We have actually thought, if we form a structure, and then use these techniques in different ways, rather than just pushing them all into one shape. So that was a breakthrough yesterday to come up with the design of the actual structure.'

-Sheldon Marie Stansfield, Design and Craft Undergraduate student

'There are little seeds of kind of possibilities that can only happen if you are put in those positions or meet those people, who are looking at things from a different perspective'

-Clare Evans, Design and Craft Undergraduate student

'The trip has provided a greater sense of scope to help me understand how the idea of sustainability is multi- faceted and may be applied in different ways. It was refreshing to try to understand the problems of sustainability from the perspective of these manufacturers, without any kind of 'green-washed' or misinformed ideas. The trip was a valuable opportunity for me to build upon some aspects of my earlier research and could lead to future initiatives based on what I came away with from the trip. This has helped bridge a gap between my understanding of some of the real world drivers for change and a move towards sustainability'.

-James Dart, Design and Craft Undergraduate student

This concept is called 'flipping the classroom[3]', where students learn from one another, and off-site in a peer learning community and experience accelerated learning on a voluntary basis, without being restricted by marks, scores or module results, which allowed complete immersion in the research project.

By exploring a live teaching model with university partners, which was supported by EU-funded research and, therefore, not a drain on institutional budgets and resources, high level engagement in sustainable theory and practice with undergraduates, postgraduates and staff was facilitated. This engagement was all outside of the standard curriculum activities and led to informing and being subsumed into the curriculum and into the core of student work and philosophy.

2.3 BRIDGE exchanges and fieldwork

As a result of facilitated cross-border knowledge exchanges between France and the UK, where participants explained their research and commercial thinking, the BRIDGE project partners were able to challenge their understanding of the sustainable agenda (people, profit and planet) and re-assess their position and future direction from a more informed perspective. The exchanges between the UK and the French commercial and educational communities challenged and invigorated the business and research teams through many eco events focusing on lifecycle analysis, production and disposal methods; concepts around 'make local', 'supply local', sell local' and 'dispose of local', as opposed to a global manufacturing supply chain process. These concepts led to heated discussions between designers, engineers, entrepreneurs and scientists about material value and environmental benefits versus simply price and a true cost calculation of goods.

Bridge Club members and partners saw an example of sustainable business practice on a field trip to a fibre producer and processor where the function of the study was to experience sustainable green philosophy and zero waste processes in action and in so doing learn from the experience as illustrated in Fig. 2. The BRIDGE community of learners saw at Diamond Fibres mill in East Sussex, a profitable SME considered to be a fine example of green economy innovation and local entrepreneurship. Diamond Fibres use old and new technology to take and process wool from small local sheep flocks (such as the ancient and rare breed of sheep Black Wensleydale Longwool). Hand-spinners, felters and fibre artists and textile designers use the wool in the natural colour ranges from the sheep fleece. The business provides a bespoke high-quality fibre processing service of carding, spinning, and yarn-plying for products which have a traceable and sustainable narrative using minimal chemicals which is increasingly important for a green consumer. The CEO of the business explained the use of different eco processes employed in turning raw material fleece, into yarn, where the end product colour range is a blend of fibres without dyes and minimal chemical intervention. Any discarded wool fibres, which are brittle, from the mill, are used in local agriculture as soil conditioners or as insulation. The function of this (and other project field trips) was to experience and learn from a successful SME who could illustrate the process of sustainable supply chain cradle-to-cradle lifecycle production, from animal breeder to the final end product. This was a scalable business which had created a unique selling point, market differentiation and jobs in a rural location.

Figure 2. Diamond Fibres Ltd. (© June 2013, with kind permission of Carolyn Watt, University of Portsmouth, UK)

2.4 Communicating research

In order to communicate the research and challenge existing business and education practices, it was imperative to communicate with as many people as possible throughout the project and so, from the outset, BRIDGE was underpinned by a series of carefully devised UK and French events that presented and explained the results to the outside world, rather than keeping the research within the silo of the institutions involved. Here, we report specifically on some of the UK research communication events and not on those which ran in tandem in France where the French events focussed upon employment, business start-up and incubation rather than material sustainability and process are being reported in separately.

The Brighton BRIDGE Circus conference and The True Value of Materials exhibition

The Brighton BRIDGE Circus conference was an information exchange and knowledge transfer event, which was held over 2 days in Brighton in December 2013. To mark the interim point of the project, the academic conference was supported by a gallery exhibition 'The True Value of Materials' that showcased sustainable products and materials from emerging creative green entrepreneurs. The sustainable agenda was explored by theoreticians and makers, who produced scholarly literature and/or physical products encapsulating the thinking and narratives for the event for all those interested. The conference's speakers included scientists, engineers, designers, practitioners, academics and experts in D-STEM-B fields and all presentations were simultaneously translated into French. The structure and agenda highlighted the diversity of 'green' perspectives, knowledge and experience on the issue of sustainability and showcased the opportunities presented by transdisciplinary relationships. Participants were invited to be provocative, and the presentations generated a number of discussions. Data analysis showed that there were approximately 300 delegates in attendance at the conference who used social networks to communicate the event content in the UK and France.

The True Value of Materials exhibition saw more than 900 visitors to the gallery, which was also seen through the gallery windows by passersby in vehicles and on foot in the city. The show ran for 2 weeks and had 30 exhibitors, including businesses, artists, designers, scientists, academics and alumni from the UK and France (Fig. 3). The intention of the creative product exhibition was to communicate the complex theories and research of the material agenda and to showcase work produced by people who understood the ethos and championed the philosophy of sustainability. In addition to the exhibition, workshops took place showcasing a range of textiles and production activities where more than 40 people participated. The Brighton BRIDGE Circus celebrated *a brand new green*: a green that showed new designs, new materials, new potentials and new uses to celebrate the future green entrepreneur and design for sustainability. The curator was careful to select and exhibit a range of sustainable product outcomes and proof of concept pieces such as sports equipment made from knitted and woven flax, items composed of biodegradable resins and new conceptual garments made from wool combined with plastic waste. Transdisciplinary ideas were shared and new territories of investigation were enabled such as a development of a milk protein thread, where designers, scientists and entrepreneurs gathered to explore concepts, analyse possibilities, develop technology and finance research using a collaborative research approach. Some of the exhibits represented the simple, local use of natural materials; others were provocations, concepts, symbols, works of the imagination visualizing data and examples of the mutability of materials and their impact as drivers of sustainable thinking for the green entrepreneur.

Figure 3. The True Value of Materials exhibition (© December 2013, with kind permission of Exploding Sky Images, UK)

Ultimately, BRIDGE established a green web between regions of France and the UK as a step towards creating employment using sustainability as the core driver of progress

2.5 Knowledge Transfer Network (KTN) Materials Research Exchange, 2014

The importance of the BRIDGE project was recognized by a UK government invitation from the Technology Strategy Board (TSB) to exhibit at the prestigious Knowledge Transfer Network (KTN) Materials Research Exchange, which took place in Coventry, UK, in 2014 and which the TSB organize. The event attracted over 500 commercial and research visitors, the majority of attendees from universities such as Oxford, Brunel and Southampton and the private sector, including Jaguar Land Rover, GKN Aerospace and Morgan Advanced Materials.

This is a prestigious exhibition of leading research institutes in the science, technology, engineering and mathematics (STEM) disciplines. The aim and scope of the exhibition was to showcase the richness of UK materials research, bring closer collaboration between the industry and the UK materials research base, increase public and private funding, and private equity support for commercialization. The show provided an opportunity for KTN to facilitate the coming together of all the finest material research in the UK through exhibition stands, seminars, workshops, keynotes from government ministers and presentations of student research posters. This event enabled BRIDGE to present work from a sustainable design and production, materials and creative industry perspective.

2.6 The Eco Technology Show, 2014

The Eco Technology Show, organized by an esteemed advisory board, was held at the Brighton Centre and took place in 2014. The show exhibited low carbon technologies across the built environment, energy, transport and resource efficiency sectors and offered a forum for networking opportunities through exhibiting stands, communication hubs, keynotes, panels and talks. In total, there were 136 exhibitors and 70 keynotes, panels and talks during the show. The Eco Technology Show was an opportunity for BRIDGE partners to network with businesses.

The Eco Technology Show had 3500 visitors over 2 days: 26 % general public and 74 % trade (18 % public sector, 35 % private sector and 10 % building and property professionals). Visitors came from the south of England, Scotland, Wales, Northern Ireland and Ireland and other countries in Europe.

To communicate the project widely, BRIDGE hosted a 1-day symposium titled 'The True Value of Materials: Waste' on the last day of the Eco Technology Show. The partners in the project from the UK and France all participated in, and attended, the symposium; which brought speakers' together to consider the following question: 'Is waste the new raw material?' Again, this provided a platform for discussion of transdisciplinarity and sustainable materiality (Fig. 4).

Figure 4. BRIDGE stand at the Eco Technology Show, 2014 (© June 2014, with kind permission of Ashley Davison, LikLife Images, UK)

Quantitative data was collected from these showcases; however, at the time, it was unclear what impact these events would have on our community of learners. Eventually, from the qualitative questionnaire data and quantitative data on visitors and their businesses, it became apparent that there is a wide interest in sustainable materials and a will to engage with production and consumption from all those involved.

3. DISCUSSION AND EVALUATION

3.1 Visualizing research: a series of examples from the BRIDGE project

BRIDGE led directly to the establishment of new entrepreneurs through their engagement with sustainable material values, learned during their involvement with the partners and activities throughout the 3-year project term. Sustainable principles, using local eco materials and traditional light touch production techniques, informed their processes through interdisciplinary knowledge exchange. A selection of UK green entrepreneurs are acknowledged here as narrative case studies because the project directly impacted on their work, either via providing funding or providing insight towards new learning, research and development and commercial opportunities.

James Dart, 3D designer (Master's degree at École Cantonale D'Art de Lausanne (ECAL) in Switzerland), focused his research on new sustainable material knowledge using flax and knitted formed structures (Fig. 5). His knowledge and rationale for using degradable bio-resins for sport accessories came directly from his engagement with BRIDGE and bio-resin producers in the UK and his understanding of the environmental crisis of the use of non-biodegradable composites in products, something at odds with the wellbeing messages from the sports industry.

Clare Evans (Master of Sustainable Design, UoB) considered fashion and textile waste (Fig. 6). The concept behind her project was to use raw material waste in a new material for contemporary applications. Textile waste from French partner ESITPA that was being repurposed into insulation blocks used for walls was combined with corn starch plastic, heat pressed into sheet material and then moulded into contemporary luggage. The products are visually intriguing and the consumer's wish is to know more of the narrative.

Figure 5. James Dart (Duo Lin): BRIDGE p was showcased as part of the BRIDGE Circus, KTN Coventry event and Waste event and has gone on to study at ECAL at Masters level (© December 2013, with kind permission of James Dart, ECAL, Switzerland)

Toni Hicks (textiles designer and Senior Lecturer in Textiles at UoB) and her team held 'The Continuum' workshop to demonstrate textile processes, which allowed participants to experiment with different material properties and techniques to improve sustainable manufacture (Fig. 7). Needle punch felting, wet felting, knit, heat pressing, lamination and laser cutting were used on wool and flax from sustainable sources in the south of England and northern France, and waste materials, including plastic bags, hair, dog fur and beeswax, were incorporated. The materials were selected because of their communicative properties regarding the environment and ethics. Visitors to The True Value of Materials gallery were encouraged to touch the pieces and reflect and comment upon the issues surrounding the sustainable agenda. Books of provocative pertinent comments are an additional record of the project impact on consumers.

Figure 6. Clare Evans, The Fluff and Nonsense project (© June 2014, with kind permission of Ashley Davison, LikeLife Images, UK)

Kirsty McDougall, designer at Dashing Tweeds (Senior Tutor in Woven Textiles at the Royal College of Art), exhibited Lumatwill™, a fashion fabric that combines reflective filament with traditional worsted or tweed wool yarn that has a recycled content (Fig. 8). The cloth appears to be a traditional man's suiting or tweed but is reflective in certain light and would suit the urban cyclist or pedestrian. The company uses British tweed textiles and workshops to create traceable durable fabrics and garments as part of a sustainable philosophy, which has a brand identity and a niche clientele with significant disposable income.

Jasmine Nicholls, a textile designer, was interested in sustainability within the textile industry using local organic material suppliers and multimedia as a marketing tool. Nicholls' business, 'Rudimentary', employs a responsible design ethic, ideals of utility and simplicity in fabricating desirable yet durable products (Fig. 9). Her distinct uses of waste materials have been manipulated and constructed to convey a sophisticated approach to recycling textiles. Through the BRIDGE project, valuable contacts with local practitioners, suppliers and retailers were made and work exhibited in all events.

Figure 7. The Continuum: a group of staff and students worked on 'The Continuum' demonstrating some of the properties that wool and flax possess when combined in different techniques. The final product was exhibited at the Brighton BRIDGE Circus (© December 2013, with kind permission of Exploding Sky Images, UK)

3.2 Case study exposure; online presence and dissemination

MIRIADE presented the first iteration of the BRIDGE channel eco-cluster website at the official launch of BRIDGE in Caen in 2012. Further to input and feedback by all the partners, the website was officially launched at the Eco-Design event in Alençon. This web platform[4] documented BRIDGE events and activities and, in July 2013, provided an online platform through which members could identify potential partners to develop innovative projects and cross-channel collaborations.

Figure 8. Lumatwill™ was created in 2005 by Kirsty McDougall and Guy Hills as part of the range of fabric offered by their woven textiles fabric business, Hills McDougall and Dashing Tweeds (© December 2013, with kind permission of Exploding Sky Images, UK)

The BRIDGE research team at UoB documented their design research scoping activity via a WordPress blog[5] created and edited by Marney Walker. A Facebook page and Twitter account (@BrightonBRIDGE), edited by Mylinh Nguyen, Carolyn Watt and Harriet Parr, were developed to share knowledge and to document design research and development activity. The blog attracted views from a global audience across 77 countries worldwide, with 4881 views from the UK, 529 views from France and 328 views from the USA as of November 2014 Fig 10.

For quantitative analysis as of 11 November 2014, the blog had received 6707 views, with 1953 visitors. Figures 11 and 12 illustrate how views and visitors during BRIDGE activities increased significantly when articles were posted leading up to and about the events. The screenshot image in Fig. 13 shows search referrals from different websites, with the highest coming from UoB's Faculty of Arts website with 398 referrals, followed by 275 referrals from Facebook and 116 referrals from Twitter. Other referrals came from people affiliated with the BRIDGE project's personal websites. The BRIDGE Eco Cluster website, led by MIRIADE, attracted 3943 visitors by February 2015 and initiated 5581 connections between academics and business in the field of sustainability.

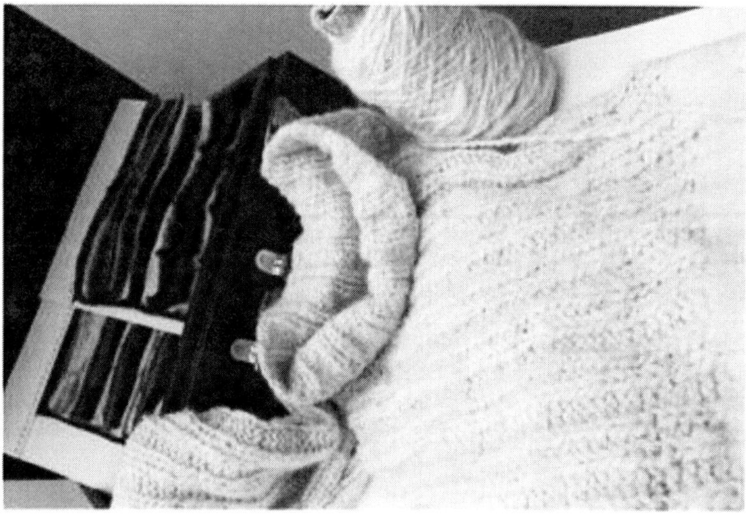

Figure 9. Jasmine Nicholls (Rudimentary): BRIDGE part funded materials for this work which was showcased work as part of the BRIDGE Circus

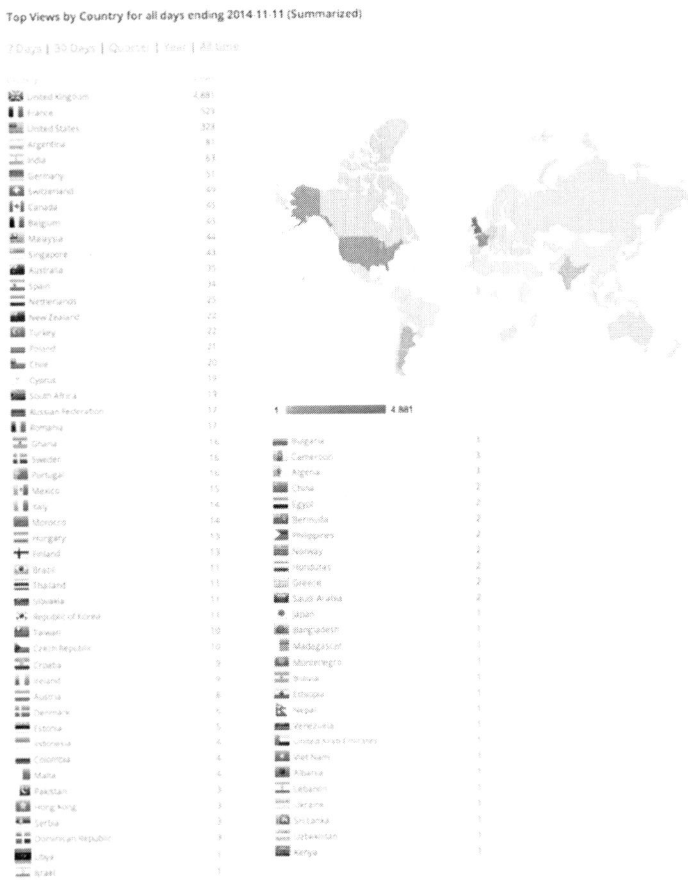

Figure 10. Screenshot image of summary views by country (© December 2014, with kind permission of Mylinh Nguyen, BRIDGE team research assistant, UK)

Figure 11. Screenshot image of the viewing statistics by month (© December 2014, with kind permission of Mylinh Nguyen, BRIDGE team research assistant, UK)

Date	Event	Views	Visitors
September 2013	The Bentley Wood Fair	470	113
December 2013	Brighton BRIDGE Circus	1,341	328
February 2014	The Green Business Start Up Launch Pad KTN Materials Research Exchange	419	89
June 2014	Eco Technology Show	279	80
September 2014	Make Lewes Festival SheSaid Brighton	321	93

Figure 12. Table showing how certain events and blog posts correlate to the viewing and visitor figures (© December 2014, with kind permission of Mylinh Nguyen, BRIDGE team research assistant, UK)

Figure 13. Screenshot: search referrals (© December 2014, with kind permission of Mylinh Nguyen, BRIDGE team research assistant, UK)

4. CONCLUSIONS

What we learned

4.1 Science and design: building on new connections and recognizing mutual benefits

Given the diversity of the partners, their understanding of the green economy and the distance they have travelled regarding subject knowledge, the mutual benefits for the contributors are only beginning to become apparent. The number of deliverables were ambitious yet all were achieved in the time frame, but further work is needed to build on and analyse these new connections, relationships, knowledge and methods, in order to, improve upon, communicate and consolidate the benefits of new knowledge that the BRIDGE project facilitated. Theoretical analysis of different elements and impacts of the project research is currently taking place for a range of international publication, which includes business, education and design journals. The rationale for the importance of this descriptive journal article is to publish the transdisciplinary methods narrative here, separately and urgently, in tandem with the development of other publications which are related to each other but with different focus thus enabling the whole complex project picture to be communicated and understood. In addition, the work has been presented at conferences in China, the UK and Spain. Data gathering and analysis on technological developments, research and development innovations and new startup businesses from BRIDGE is work in progress and communication of the project relating to sustainability, good practice and partner experience is broadcast as short videos on YouTube sites. A new follow-on research project building upon BRIDGE structure and methods from a sustainable job creation and social innovation point of view is in development which speaks to the following bullet points:

- Future partnerships to include other disciplines informed and underpinned by design thinking, innovation and knowledge transfer.

- Taking advantage of emerging funded research opportunities including technology materials, design science and business collaboration, as a motivator for employment.

The above two suggestions, from lessons learned in two countries, are contingent on a coordinated approach to communicate, engage and explore findings and dissemination methods in order to maximize and maintain engagement and improvement across discipline areas to the expert and non-expert. Managing a complex multi-site multi partner project such as BRIDGE with such diverse stakeholders in terms of outputs, practice, experience, location and age groups was a challenge, which needed diverse communication methods to deliver a consistent message. Now, there is a depository for data and information on the project, accessible by all partners and the public for open source information to be used as a possible building block for other initiatives for increased employment, education and better environmental processes.

This work contributes to a growing body of literature with a focus on transdisciplinary thinking within design, wellbeing and sustainability and will serve as a basis for future studies and development for externally funded research projects, the potential to influence government policy in the education and the small business sector. The new green entrepreneurs who are emerging using closed loop more sustainable research models with an appreciation of materials are one solution to the issue of overproduction, irresponsible consumption and waste, where all materials are a valuable commodity.

4.2 Growing green entrepreneurs: creating a toolkit of skills and knowledge

The long-term impact on the young entrepreneurs, who have engaged in the challenges presented by BRIDGE in sustainable design and new business practices, and how that has influenced their future career contexts and directions, will be measured over time. The project created a simple how-to toolkit for new green business start-ups, loans and grants, with pathways to mentors and expertise. This guide is intended for young entrepreneurs, researchers and engineers in France and Britain who are interested in starting a business in innovative technologies. The guide explains all the steps to take and useful contacts to establish in order to facilitate the process of creating these types of business projects in Europe. Educational institutions particularly in textile material product design courses are beginning to recognize the importance of creating opportunities to engage young people at an earlier age in sustainable business development, using research projects such as BRIDGE to showcase the potential outcomes to inform, inspire and create opportunities as they progress and thus create market opportunity and differentiation.

In terms of the authors' original insights in relation to material sustainability, design and innovation, the project findings have underlined that producers and consumers take meaning and understanding from materials and products, in order to give value to goods and these meanings should be honestly communicated. Through the many artefacts and products produced throughout the project, exhibited and discussed in the UK and France, the stakeholders began to understand the holistic narrative of and the origins, product journey and the true cost rather than price of an item. Using the transdisciplinary approach, the community of learners began to differentiate between 'good and bad' products and processes and relate their understanding of production and systems to the sustainable pillars in Brundtland's description, people, profit and planet to their understanding of honest brand identity. To address the issue of green innovation in terms of production, consumption and reuse of materials, a multi-faceted approach is necessary to develop green entrepreneurs and business for the discerning consumer, and as the project yielded information from those involved, giving information with integrity in as simple a way as possible transferred complex knowledge to the non-expert consumer. The partners concluded that the creative industry has expertise in the communication field regarding Building Research and Innovation Deals for the Green Economy and the research project has confirmed that the collaboration with Design, STEM and Business (D-STEM-B) partners true innovation in transdisciplinary and applied research towards a more sustainable future lies.

ENDNOTES

[1]University of Brighton, 2014. BRIDGE Project Website. [online] Available at: <http://arts.brighton.ac.uk/projects/bridge> (Accessed 9 May 2014) INTERREG-funded BRIDGE project was a European Union INTERREG IV-funded project, which aimed to demonstrate the opportunities for eco innovation arising from transdisciplinary research partnerships. The project linked nine partners from southern England and northern France from diverse settings: design research, sustainable construction, waste reprocessing, business incubation, agricultural engineering and sustainable materials research. BRIDGE aimed to build partnerships for cross-border economic development and complementary centres, business and design incubators, in order to share the best practice in eco design between the two countries.

[2]From the World Commission on Environment and Development's (the Brundtland Commission) report Our Common Future (Oxford: Oxford University Press, 1987). *Our Common Future*, also known as the Brundtland Report. Our Common Future United Nations 1987 UN Documents: Gathering a Body of Global Agreements has been compiled by the NGO Committee on Education of the Conference of NGOs from United Nations websites with the invaluable help of information and communications technology.

[3]Educators Technology (2013) Website http://www.educatorstechnology.com/2013/09/flipping-classroom-simply-explained.html. Accessed 13 July 2015.

[4]BRIDGE Eco Cluster (2014) Website. http://en.bridge-ecocluster.eu. Accessed 13 July 2015.

[5]BRIDGE Brighton (2014) Website. http://bridgebrighton.wordpress.com. Accessed 13 July 2015.

ACKNOWLEDGEMENTS

This work is supported by the INTERREG IV A France (Channel)—English Programme.

Partners include MIRIADE, ESITCaen, Esitpa, Medway Council, Normandie Incubation, Orne Developpement, Remade South East, University of Brighton and WSX Enterprise Ltd.

Contributors include Mylinh Nguyen, Harriet Parr and Tom West.

COMPETING INTERESTS

The authors declare that they have no competing interests.

AUTHORS' CONTRIBUTIONS

JF, lead author, was the Principal Investigator of the INTERREG IV BRIDGE project leading the University of Brighton, UK. JF was appointed in January 2015 to the University of Portsmouth Faculty of Creative and Cultural Industries as Associate Dean of Enterprise and Innovation and Professor of Design and Innovation. All authors, JF and CW contributed to the BRIDGE project, participated in the design, coordination and drafting of the manuscript. All authors read and approved the final manuscript.

REFERENCES

1. Braungart, M, & McDonough, W. (2009). *Cradle to cradle. Remaking the way we make things*. London: Vintage.

2. European Commission (2015) Social Investment. http://ec.europa.eu/social/main.jsp?catId=1044&langId=en. Accessed 11 August 2015

3. Mills, R. (2014). The value of networks in bringing materials research and innovation to market. *Translational Materials Research, 1*, 1. Available through: IOP Publishing Ltd website < http://iopscience.iop.org/2053-1613/1/2/020202> [Accessed 10 December 2014].CrossRef

4. Piaget, J. (1972). The Epistemology of interdisciplinary relationships. In Apostel et al. (Eds.), *Interdisciplinarity. Problems of teaching and research in universities* (pp. 127–139). Paris: OECD.

5. RCUK. (2011b). *Research Councils UK Strategic Vision* (p. 8). Swindon: RCUK.

6. University of Brighton. (2014). C-Change Website. http://about.brighton.ac.uk/sustainability/campus/cchange/. Accessed 17 December 2014

7. University of Brighton. (2014). DR-i, Design Research Initiatives Website. http://arts.brighton.ac.uk/research/design-research-brighton/about-dr-i. Accessed 30 November 2014

8. University of Brighton. (2014). Green Growth Platform Website. http://sustainablebusiness.org.uk/network/green-growth-platform/. Accessed 18 December 2014

9. University of Brighton. (2014). Sustainability Network Website. http://arts.brighton.ac.uk/research/sustainability-network. Accessed 18 December 2014

10. University of Brighton. (2014). The Waste House Website. http://arts.brighton.ac.uk/business-and-community/the-house-that-kevin-built. Accessed 17 December 2014

3

ETHICAL FOUNDATIONS IN SUSTAINABLE FASHION

Kirsi Niinimäki

Design department, School of Arts, Design and Architecture, Aalto University, PO Box 31000, FI-00076 Aalto, Finland

ABSTRACT

Sustainability is fuzzy and wide concept and the discussion what to sustain continues, the resources or lifestyle. Furthermore how holistically the sustainability should be approached is under discussion. According to the holistic approach for environmental ethics ecosystems and biosphere as a whole should be considered, not individual's rights. Furthermore in ethical discussion the value aspect is most important; what is considered to be valuable and from where value comes from. Furthermore environmental ethics raises deep questions; who counts morally and why. Moreover the human actions are under evaluation in environmental ethical discussion; "how should human beings act in the nonhuman natural world".

Environmental ethics can be applied in other fields e.g. in design and fashion. Values and ethics are fundamental grounds also for sustainable fashion. Environmental pioneer 1995 argues that "ethics are the philosophical basis for making choices about morals and values". He further continues that "to think dispassionately about what we design and why, and what the eventual consequences of our design intervention may be, is the basis of ethical thinking". Accordingly it is worthwhile to investigate the value base in sustainable fashion and further to consider the consequences of our design and industrial manufacturing processes in the clothing sector based on environmental ethics. Moreover it is important to understand the consequences of our design practices; e.g. how current design, manufacturing and business practices affect unsustainable consumption patterns. Products actually configure consumers' needs and use patterns and hence design can be said to be "practice-oriented", which leads to certain everyday consumption habits. Therefore designers, manufacturers and companies are responsible not only for the environmental impacts of the fashion industry but also for the unsustainable consumption behaviour of consumers and the increase of waste streams, i.e. the unbalance in the fashion system.

This text provides an overview for ethical foundations in the fashion field. Even though a lot of research has been done lately in the sustainable fashion field, the views have been limited and not well grounded on philosophical knowledge about environmental ethics and value discussion.

KEYWORDS

Environmental ethics Values, Sustainable fashion, Corporate social responsibility, (CSR) Ethical consumption, Extended producer responsibility (EPR), Sustainable business Greenwashing

1. INTRODUCTION

Sustainability is fuzzy and wide concept and the discussion what to sustain continues, the resources or lifestyle. Furthermore how holistically the sustainability should be approached and understood is under discussion (e.g. Scherer 2003, 355). According to the holistic approach for environmental ethics ecosystems and biosphere as a whole should be considered, not individual's rights (Palmer 2003). Furthermore in ethical discussion the value aspect is most important;

what is considered to be valuable and from where value comes from (*ibid.*). In addition environmental ethics raises deep questions; who counts morally and why. According to Palmer (2003) firstly the community, rather than individual should be in the focus while evaluating moral significance. Secondly ecological qualities have primary value. Moreover the human actions are under evaluation in environmental ethical discussion; "how should human beings act in the nonhuman natural world" (*ibid.* 17).

This text provides an overview for ethical foundations in the fashion field. Even though a lot of research has been done lately in the sustainable fashion field, the views and research approaches have been limited and not well grounded on philosophical knowledge about environmental ethics and value discussion. Interesting research is emerging mainly in the small niche areas about sustainable fashion, its meaning and connection to use phase and individual consumer's needs and participation (e.g. Fletcher 2008; Fletcher and Grose2012), but in fore mentioned the focus has been mainly on a small and design-driven initiatives. This text focuses on the change in the industrial scale.

This text is based on author's earlier writings extended with other experts' views of values in the sustainable fashion field. This knowledge is reflected against environmental ethics. The text starts by reviewing environmental ethics and the current discussion going on in the sustainable fashion field. Thereafter new business thinking in the context of sustainability, innovation for sustainable business, future oriented thinking, collaboration with the customer, CSR, EPR, animal ethics and ethical luxury is presented. Moreover the text discusses about greenwashing and ethical consumption. The text ends with the consideration how to design value.

2. REVIEW

Environmental ethics can be applied in other fields (Light and Rolston III 2003) e.g. in design and fashion. Values and ethics are fundamental grounds for sustainable design. Environmental pioneer Viktor Papanek

(1995, 70) argues that "ethics are the philosophical basis for making choices about morals and values". He further continues that "to think dispassionately about what we design and why, and what the eventual consequences of our design intervention may be, is the basis of ethical thinking" (*ibid.*). Therefore ethical thinking takes account the moral implications of our actions (Clegg 2011). Accordingly it is worthwhile to investigate the value base in sustainable fashion and further to consider the consequences of our design and industrial manufacturing processes have in the context of sustainable development.

It would be fruitful for designers to consider whose values are included in the design process and on what basis. It is also good to remember that there isn't any value-free design. We should design and construct a new system and basis for value creation, to help us face future challenges in the fashion discipline. Such a system could serve to evaluate design proposals and their benefits for sustainable development in a long term. Furthermore manufacturing processes, business models and marketing ethics could be included in this value evaluation process. However we need novel knowledge on how to embed the dimension of environmental value and future-oriented thinking into the current fashion system.

While including the ethical dimensions in the sustainability discussion, we have to start the discussion from values. Values relate to beliefs about good and bad, right and wrong, and they are the basis that gives direction to our moral behavior (e.g. Gibbins and Reimar 1999). Paehike (2000) has listed the core values in environmentalism: firstly, the protection of biodiversity and ecological systems; secondly, consideration of negative impacts on human health; and finally the sustainable use of resources. Moreover he argues that these values are not sufficient, and we must challenge our whole way of living: how we organize our societies and how we live our lives, including new challenges in purchasing and use patterns and in disposal behavior (*ibid.*).

Leiserowitz, Kates and Parris (2006) argue that the most important value in sustainable development is environmental protection. This is a very broad concept and includes many different values and approaches. In addition to environmental protection, in sustainable development there are three key 'stakeholders': people, the economy and society, whose sustainable development also needs to be taken care of in the quest to achieve a sustainable balance in the system. In this process values such as equality, freedom, and democracy are most important (*ibid.*).

Walker (2007, 70) argues that in the context of sustainable development it is possible to add ideological value to objects. This means that through environmental values we can assess products in a completely new way. Through this lens it is possible to value e.g. redesigned objects according to their environmental value. Accordingly eco-materials, leftover materials, recycled products and parts, and a 'classical' appearance can be evaluated with the goal of increasing the environmental and sustainable value of the product. Moreover renting, leasing and other actions aiming to dematerialize consumption can be seen as sustainable value actions (Niinimäki 2011).

The ethical and value discussion is complex and hard to do in the industrial context. As today it seems that possibilities in industry to make ecological and ethical choices are limited, designers and producers do what they can, a situation that can be described as realistic thinking. This kind of value-based, yet realistic but narrow, approach to sustainable fashion is the most common approach currently existing in the business. Best practice can be defined to mean that designers and companies choose the best environmental and social solution that exists (Niinimäki 2013). Especially small- and medium-sized companies have to make their choices from limited possibilities. Finding suitable eco-materials that can be ordered in small amounts, to give only one example, can be most challenging for small companies (Niinimäki and Aakko 2014). What an entrepreneur can do is thus limited and defined by the resources available as well as based on the current rather narrow knowledge base. Accordingly available resources create the action framework for sustainable-oriented designers and companies: i.e. what kinds of design and production choices they can make. The bigger the fashion company, the more power and possibilities they have to influence sustainable practices in the field. Despite these limitations, it is important that designers and companies identify their own value base, act accordingly, and inform their customers about their values and ethical actions (Niinimäki2013).

2.1 New business thinking in the context of sustainability

New radical business thinking is needed to create a bigger change in the fashion field. The economic and industrial systems of the fashion industry are currently based on extremely fast cycles of production, fast-changing trends and planned obsolescence of the products. As garment prices fall, producers must maintain profitability by increasing the amount of sold units, which has led to a huge amount of fashion products on the market and extremely saturated markets (Niinimäki 2011; 2013). We must find new radical ways to create a win-win situation for both consumers and manufacturers - for all stakeholders - and for sustainable development. We need more knowledge about consumers and the consumption side to create a sustainable transformation process inside the fashion industry and business that leads to sustainable consumption practices. We also need to develop foresight and new systemic thinking on how to move towards sustainable societies, in order to see sustainable practices as an opportunity to design and not to think that sustainability is an obstacle to doing business. Very often, the sustainable business approach is feared to mean de-growth, but it can also be seen as a new challenge and opportunity to do things creatively and differently (Niinimäki 2013). Sustainable business logic is based on fundamentally different logic than traditional fashion business and not on the growth logic alone (Fletcher and Grose 2012).

Sustainable values lay the ground for credible and responsible sustainable business and marketing. Sustainable business and marketing includes not only a sustainable value base but also ethical consideration. This aspect seems to be problematic for some manufacturers and companies, they do not want to make moral claims about what is right and what is wrong. However, for strategic sustainability thinking, it is important to consider not only what efforts and investments to make but if would it be worthwhile to withdraw from some markets because of ethical reasons or too big social or ecological risks (Belz and Peattie 2011). In this regard, a deep assessment should be made according to the company's own value base. Including sustainability in the company's strategic planning might open new strategic possibilities but it can also be done to prevent risks in legal, resource, environmental, reputational or sociopolitical levels (Lowitt 2011).

As the fashion industry is international and global, principles of corporate social responsibility (CSR) should be followed. Moreover, it is important to remember that consumers' trust can only be achieved through open communication and information and only through real commitment to sustainability at all levels of the company's practices, avoiding too limited a sustainable approach and a 'greenwashing' attitude. In addition, because of the global character of the fashion business, the attribute 'Made in' is today less relevant than the attribute 'Made by', which needs a lot of information gathering about the different stages in manufacturing and communicating these aspects not only to consumers but also to all stakeholders.

Stakeholders in sustainable marketing thinking are not only consumers and investors but can also include employees, business partners, suppliers, competitors, government (through regulations), NGOs, pressure groups and communities. Working with these stakeholders needs consideration of not only exchanges but also long-term relationships and future possibilities through radical sustainable innovations (Belz and Peattie2011).

The global fashion business is fiercely competitive, and simultaneously, Western consumers' environmental interest is increasing; in this environment, sustainable business thinking and sustainable marketing offer new business opportunities for the fashion industry. As it is challenging to ensure profit with the existing fashion system and current price levels, it is worthwhile to challenge the current unsustainable business models in fashion and rethink value creation from a wider perspective: value proposals from the consumers' side as well as from the environmental perspective and through creating long-term relationships between business and its stakeholders (Niinimäki 2013). In this context, it is possible to create sustainable value proposals and stimulate new business thinking and business models that contribute to a decrease in materialistic consumption.

2.2 Innovation for sustainable business

Vezzoli (2007) argues that radical innovations are needed which do not only depend on technological development and innovations but also stimulate new interaction and partnerships between different stakeholders as well as new sustainable relationships between consumers and products. If we want to change the fashion system towards a new kind of balance, we need new kinds of radical design and business thinking. We have to change the current economic paradigm, and we have to do business differently than we do today. Business and industry have to redefine their tasks more radically in order to usher in significant eco-innovations and most importantly all-new business strategies (Ottoman 2011, 90). As Kemp (2008) reports, a fundamental change is needed to reach systemic change, but transition faces resistance and it takes time to emerge. On the other hand, system level changes can be based on paradigms, understanding how ideas relate to each other to form accepted models (Fletcher 2008, 73, based on Meadows 1997). As Fletcher (2008, 71) argues 'if we influence things at the level of a paradigm, then a system can be totally transformed'. This system level transformation needs questioning current practices and understanding, and further, it needs a new way of seeing and understanding.

When we talk about eco-innovations, we mean products, services or processes that offer value not only for both the business and customer but also to the environment through significant decrease in environmental impact. Sustainable innovation is a larger systemic approach where sustainability thinking is integrated in all levels of the business (products, technologies, services, new business models, organisation model and relationship with stakeholders) (Niinimäki 2013).

The levels of innovation are described as follows (based on Brezet and van Hemel 1997):

(1)Incremental (small improvements for existing products)

(2)Redesign (major redesign for existing products)

*(3)*Product alternatives (new product or service concepts)

*(4)*System innovation and design for a sustainable society

The first level of innovation refers to small, incremental product improvements. The second means redesigning existing products and production processes (e.g. in an eco-efficient way), and the third innovation type aims to find alternative methods of delivering the same function (Brezet and van Hemel1997). Currently, our efforts in sustainable fashion are between the second and third levels. However, these innovation types are not enough, and we need to move towards more radical and systematic innovative thinking while challenging the whole industry and its current practices of doing business.

Tischner and Charter (2001) identify four approaches to sustainable design: repair, refine, redesign and rethink. The fourth approach is the one we should aim for and close to the forementioned system level innovation. A *rethink* requires a radical change in our mind set, and it can offer breakthroughs for new lifestyles, ways of living and doing things, as well as approaches to fulfil consumer needs in a more sustainable manner (Tischner and Charter 2001). This approach needs strategic innovations that lead to new business models. As Tukker et al. (2008) argue, these new practices often stay in niches for a long period until a window of opportunity opens and their breakthrough is possible.

For this radical and strategic innovation approach, we need a much more creative way of conducting the business of fashion. Strategic innovation questions who the customer is, what products or services should be offered and how to offer those products and services (Markides 1997). It is not merely about rethinking the fundamentals on the supply side but also about redesigning the business on the demand side, e.g. in the form of the user experience and rethinking value creation moving from product design to systems thinking (e.g. designing product-service systems) or designing products for closed loop system.

Several companies acknowledge that market competition is driven by products' *meanings*: 'why' people need a product rather than 'what' they need in a product. As Verganti (2009) argues, people use things for profound emotional, psychological and sociocultural reasons as well as utilitarian ones, and this understanding is most important in creating successful sustainable innovations for business. This deep understanding of consumers also lays the ground for transforming business models and creating for example sustainable product-service systems to fulfil the actual needs behind product use in a less materialistic way (e.g. Armstrong et al. 2015).

Innovations for sustainable business in the textile and fashion sector can be driven by technological improvements or technological innovations (like digital technology), economic changes (new sustainable business thinking and a systems approach) or legislation. With regard to legislation, new regulations for extended producer responsibility (EPR), for example, can change the logic in the fashion system, and preventing and/or recycling waste might be one future business model if EPR begins to cover the textile and fashion business. Innovation can also be driven by more abstract value issues, e.g. safeguarding the corporate or brand value from a bad reputation.

Succeeding in innovation in the sustainability context always needs a new kind of business logic. Driving sustainable innovation, whether to a more radical or incremental path, needs experimental and creative thinking, and this experimentation can get support from lead users before presenting these ideas to all customers. Furthermore, it is important to create networks of sustainable-oriented companies. These networks can support a company's own values-based practices and offer help in reaching their goals.

2.3 Future-oriented thinking and collaboration with customers

New kinds of relationships with the customer create new kinds of value in the business. Through a deeper relationship with the customer, a company can create a long-term dialogue with the end-user and through this dialogue create new business opportunities, new understanding of the customer's true needs and desires and robust understanding of the customer's readiness to change consumption habits. Simultaneously, the company can communicate its values and practices to customers and thereby create trust and customer loyalty towards the brand.

Through this approach, the company's focus moves from a single transaction and short-term economic thinking to long-term sustainable strategic planning and future-oriented thinking. Customer satisfaction and customer value will be at the core of the company's practices. Traditional marketing focuses on profit and revenues. Sustainable marketing also concentrates on psychographic aspects like consumer satisfaction and consumers' interest to stay loyal to the brand (Belz and Peattie 2011).

New business approaches are possible, but they require radical business thinking, focus on the consumer side and the garment's use phase and, most of all, consumer satisfaction issues. According to Manzini (1994), to change the system towards sustainability, it is not enough to redesign existing products and make some eco-efficiency improvements in manufacturing processes if the aim is to reach a more sustainable future. Eco-efficiency aims to decrease the environmental impact of industry while simultaneously using less material while manufacturing more products and resulting in less waste. The eco-efficiency approach also means savings in material, energy and chemical costs and waste disposal costs (Niinimäki 2011).

On the other hand, by taking consumer-based eco-efficiency into account, it is possible to open up opportunities for the fashion business that aim for sustainable consumption patterns. Hence, consumer-based eco-efficiency thinking in business can transform the fashion system towards re-directive practices that aim for deeper consumer satisfaction and therefore sustainable consumption patterns. Consumer-based eco-efficiency focuses not only on the environmental impact of the use phase but also on product quality and consumer satisfaction issues (Park and Tahara 2008; Niinimäki 2014). Through the satisfaction approach, value can be created for the consumer, the company and even the environment. Satisfied customers use products longer, and this is an opportunity to decrease the amount of purchased clothing and slow down consumption. This is a value opportunity from the environmental viewpoint. Satisfied consumers are loyal customers, and they repeat their purchase behaviour and continue buying brands that can provide them with product satisfaction on many levels, and this is a value opportunity for sustainable business (Niinimäki 2014).

2.4 Ethicality in production and CSR

In today's reality in the fashion field, more and more garments are manufactured in other countries, very often on the other side of the globe. About 80% of clothing exports are shipped from undeveloped countries to developed economies. For example, in Finland, about 90% of sold garments are imported from Asian and Far East countries, about 5% from European Union (EU) and other countries, and only about 5% are manufactured in Finland. In this fragmented and globalised supply network, the main issue is risk management. From the recent accidents and fires in Bangladeshi textile factories, we know that the lack of awareness about the weak points in the supply chain creates a reputational risk, at the very least. CSR tackles these issues: i.e. what enterprises' responsibilities are regarding their impacts on society. Additionally, corporations should integrate social, environmental, ethical human rights and consumer concerns into their strategy in cooperation with their stakeholders. Corporate social responsibility is used in parallel with the terms corporate citizenship, corporate responsibility and corporate sustainability (Niinimäki2013).

If the producer wants to address sustainability issues in the supply chain to a greater extent, it is important to select subcontractors well and demand good practices. The code of conduct principles help companies in this process, and standards like SA 8000, ISO 14001 and Eco-Management and Audit Scheme (EMAS) help when considering environmental and social aspects in design and manufacturing practices. Additionally, the UN has defined ten principles called the 'Global Compact' to set standards for subcontractors (https://www.unglobalcompact.org/abouttheGC/thetenprinciples/index.html). These principles take account of issues from human rights, labour issues, environment initiatives and issues in anti-corruption. Several companies have voluntarily accepted these principles and follow them in their own subcontracting arrangements, for example, Puma, GAP and

Nike. Checklists exist where producers can pick requirements for their subcontractors (e.g. http://www.csrcompass.com/).

Better than just giving the code of conduct list to manufacturers is to work in collaboration to improve the situation. Working with subcontractors and establishing a tight relationship, mutual understanding and trust take time. The producer's goals and requirements in sustainability have to be well communicated and preferably based on mutual motivation. Furthermore, to build motivation, it is uplifting to show successful examples of how to work with sustainability in the fashion field (Kruger et al. 2012).

Companies can benefit and gain great advantages by using charity or donation in their CSR strategies. The charity work should be rooted in the companies' actions strategically so that it engenders long-term business opportunities and 'improves the competitive surroundings through charity...' (Porter and Kramer 2006, as cited by Kruger et al. 2012, p. 15). It is important to realise that acting responsibly and promoting CSR do not necessarily mean direct revenues to the company. The motivation to act responsibly is based on values other than economic but actions might also be (or should be) economically successful in the long run.

The sustainability strategies should also be included at all levels inside the company. All departments should commit to sustainability practices, and all processes and products should include sustainability principles. Furthermore, the sustainability of business is the most important, so the business must be profitable. It is also important to understand that sustainability and CSR needs constant evaluation; therefore, it is a constant commitment and learning process for the company (van Dyk 2008).

2.5 EPR extended producer responsibility

Recently, the approach of extended producer responsibility (EPR) has gained a lot of attention and discussion in the EU. Extended producer responsibility requires the original manufacturer or producer to take back the product after its use. In EPR, the goal is that already in the design phase producers think about and plan how they can reduce environmental impact of the product after its use. This post-consumer waste should be reused, redesigned or recycled. In this concept, companies are forced to have a waste management system and preferably take-back systems based on closed or open loop thinking, or, if not, pay waste treatment costs. Closed loop thinking means that all off-cuts, waste and products after their use time are treated inside the factory's own processes, and open loop means that leftovers are handled by some other outside partner (Niinimäki 2013; Belz and Peattie 2011). The principle is familiar in, e.g. electronic products but not yet in textiles.

EPR does not yet cover the fashion and textile field but it soon might, and this would change fashion companies' action logic, as old garments mean extra cost through waste treatment. EU policy aims for waste prevention, closing landfills and effective material recycling systems already in year 2016. This would also cover textiles and fashion. Right now, textile waste ends up in landfills or as energy waste in most European countries. France is one exception. There producers, distributors and importers of clothes, linen and footwear have had to take back old products since 2008. A company has to have a take-back system itself or it can join a scheme, which is accredited by the French government, and pay into it (right now EcoTLC) (Tojo et al. 2012).

Extended producer responsibility is a good example how to transform the business and industry and how to include the environmental values in the business logic. It is an example of environmental ethics: how we should take responsibility of our actions and take account the consequences of our industrial manufacturing and business logic.

2.6 Animal ethics and ethical luxury

Animal ethics are largely discussed in the fashion field, yet the issue is controversial. This discussion needs philosophical foundation. The moral discussion, whether it is morally right or wrong to mistreat animals, has a long history in environmental ethics, longer than any other ethical discussion. Already, Immanuel Kant and St. Thomas Aquinas in the eighteenth century argued that it is morally wrong to harm animals, simultaneously arguing that animals deserve moral recognition (Light and Rolston 2003). Contemporary philosophy 'Animal Rights View Ethics' (also known as intentional ethics) points out that each individual (human or animal) has rights that should not be violated, and therefore, 'living creatures must never be used exclusively as a means to others' satisfaction' (Kruger et al. 2012,133). On the other hand, 'necessary ethics' or 'utilitarian ethics' emphasises that there has to be a balance in the sum of individual's utility/happiness/quality of life, and therefore, animals can be used in industrial production but the animals' living situation has to be evaluated based on the aforementioned balance also in the industrial production (*ibid.*). In the code of conduct by the Trade Association Dansk Fashion and Textile is stated following: 'in businesses where animals are used in labour and/or in the production (fur, wool, etc.), such animals must be fed and treated with dignity and respect and no animal must deliberately be harmed nor exposed to pain in their lifespan' (Kruger et al. 2012, 135).

Animal products are used in wool, silk, feather, leather and fur production. Large-scale industrial production causes ethical problems through animals' living condition and, e.g. while plucking feathers from live ducks and killing silkworms during silk production. Wool is a renewable material and therefore good for the environment, but practices in the wool production cause ethical problems, e.g. 'mulesing', practised in Australia, causes unnecessary pain for sheep. In 'mulesing', some skin is removed from the sheep's tail and breech to prevent 'flystrike' (Black 2011, 136).

Some animal materials such as leather come as leftovers from food production, and this can be seen as a valuable and sustainable resource in the fashion industry. On the other hand, fur farms are considered as unethical practices in many countries, and they raise much discussion and counterreactions among stakeholders because of the animal welfare. Even though circumstances on the fur farms are improved, discussion on ethics and animal rights is continuing. Nevertheless, clothes made from fur are expensive, valuable, durable and long-lasting, and they need little maintenance during use. Very often, fur coats are passed on for generations, which make these items connect with fond memories, and emotional attachment may easily form to these kinds of special possessions. It is also possible to redesign old fur clothes with high-quality manufacturing processes so they look new. Therefore, durable and long-lasting fur is a potentially sustainable alternative (Niinimäki 2013).

WILD Concept by Marita Huurinainen offers one sustainable way to find other fur resources than farmed ones. WILD Concept's furs come from animals that have been hunted in the wilderness. In Finland, many thousands of animals need to be shot every year to maintain balance in the ecosystem, especially to protect other animal populations, and normally, these carcasses are disposed of as waste. The important issue is that no extra pain is caused to animals during the hunting (e.g. avoiding foot-hold traps). The fur used for WILD Concept is bought directly from registered Finnish hunters who hunt the animals within the quotas allowed by the government. The wild origin of the furs is assured by the Wild Finnish Fur label. Together with the Wild Finnish Fur Association (Luonnonturkisyhdistys), designer Huurinainen has developed this alternative fur concept in order to prevent hunted animal carcasses ending up as waste. It can be guaranteed that the animals used for the WILD concept have not suffered to become a clothing item. What makes WILD unique is that it offers an ethical and responsible way to wear fur, as fur of farmed animals is not used. The end result is ethical luxury (Niinimäki 2013). Ethical luxury can be defined to mean something which is produced according to sustainable and ethical principles, which has a high environmental, intrinsic and aesthetic value, it is high quality, durable and long lasting. Ethical luxury is produced in slower processes and in small quantities, ending up in unique pieces or limited edition. Its high aesthetic and monetary value makes it valuable and long lasting, something to fall in love with; maintain it well, perhaps update it and pass on for the next generation.

2.7 Ethical and transparent

Since global manufacturing systems are complex, transparency is an important issue in the sustainable fashion field. Publishing all information from the supply chain, the names and locations of the suppliers and subcontracting partners sends a positive message; and therefore, transparency can build consumers' trust. An encouraging example of this practice is the company Patagonia, which has been a leader in supply chain transparency for many years. Patagonia has set sustainability principles as a core in their strategy. The company offers not only information about all its manufacturing locations worldwide but also background information about the factories they are collaborating with, as well as general environmental and social information about global manufacturing. Consumers can find all this information in Patagonia's webpages.

In the global fashion business, it has become challenging to identify the origin of a product since production processes are fragmented into several countries, and supply chains are complex. Yet, this information is valuable for consumers, and more transparency is demanded especially in clothing manufacturing. The 'Made by' label is based in the Netherlands, and its goal is transparency in clothing manufacturing. Through 'Made by' system, it is possible to trace the manufacturer of a garment through a code and the use of the internet (Fletcher 2008, 68).

According to the principles of transparency, a company can inform its sustainable and ethical actions, yet all information has to be true and objective, even the negative aspects should be informed or which efforts have not yet ended in good results. Transparency should not be only positive marketing message, or it changes to marketing ethos only and easily turns to greenwashing.

2.8 Greenwashing

Because Western consumers' environmental interest is rising, companies may be tempted to use sustainable and environmental arguments merely to increase sales. And because environmental and sustainability issues are complex, it is also easy to mislead consumers. Companies can use sustainability as a "marketing ploy" (Goworek et al. 2013, 388) - something that can be seen as greenwashing.

To avoid greenwashing, too narrow a focus on environmental issues should be avoided and a more holistic understanding is needed. For example, if a fast-fashion company uses eco-labels on some of its products or one product line, it does not truly change the fashion system or the economic logic behind it; the company's environmental load is still huge since it produces too much stuff in extremely fast cycles. A similar example is where a fast-fashion company uses paper bags instead of plastic ones. The business logic does not change with this small detail. The company's values have to be grounded deeply in sustainability, and its principles should be included and implemented substantially at all levels and functions.

The other aspect to consider is that all information about the environmental benefits of the product has to be proven. In this regard, standardised eco-labels are much better than the company's own environmental statements. Eco-labels (certificates) are based on an independent organisation's tests and evaluation criteria and are thus more objective and reliable than the company's own information. All irrelevant or confusing information should also be avoided.

2.9 Ethical consumption

One of the most important factors in the environmental impact during the use phase is the garment's lifetime. Currently, garments are far cheaper compared to household incomes than a few decades ago (Niinimäki 2011). According to Jackson and Shaw (2009, 146), in the 1950s in the UK, 30% of a household's income went to clothing purchases; currently, that figure is 12%, with a higher amount of consumed items. At the beginning of the twenty-first century, e.g. in the UK, fashion consumption increased by one third (Allwood et al. 2006, 11), and this was a tendency in all Western countries especially between years 1995 and 2005 (Niinimäki 2011). It is also estimated that even if the total amount of textile and clothing purchasing is growing, clothing will form a decreasing proportion of households' total spending because of falling prices. Due to these low prices and households' high incomes, the consumption of extremely cheap and disposable fashion with a very short life span has increased (Jackson and Shaw 2009). Textile and clothing prices have fallen, and currently, the consumer possesses more and more impulse-buy cheap garments and low-quality textiles (Niinimäki 2011). These kinds of low-quality and cheap garments are easy to discard. Therefore, extending the life span of garments is one of the most critical issues for sustainable development.

In the UK, a study showed that almost half of people's clothes have sat in a closet without being used during the last twelve-month period. It is estimated that this means 2.4 billion items in the UK alone. And most of these unused clothes are owned by young consumers aged 25 to 34 years (Belz and Peattie 2011, p. 125). Another study from the Netherlands showed that the average piece of clothing is owned for 3 years and 5 months. During that time, it is worn for only 44 days (Uitdenbogerd 1998, as cited by Fletcher 2008). These figures show that all Western consumers pose too many clothing items and use them too seldom before disposing them.

Contrary to the current system, product durability and long-term use are prerequisites for sustainable consumption (Cooper 2005). To slow down consumption, it is important to invest in high quality and durability as well as in aesthetically ageing materials, high design and lasting style. In this regard, services that aim to extend how long garments are used offer value in the sustainable development context (Niinimäki2014). One challenge in the current system is how to design products added with services that encourage consumers to adopt a more environmentally responsible behaviour.

When focusing on ethical and sustainable use and consumption, the following issues should be considered (Niinimäki 2013):

–Purchasing fewer garments

–Investing in meaningful garments (promoting emotional bonding)

–Investing in durable garments, more classical style and high quality

–Investing in eco-materials and eco-labels

–Extending garments' owning time and using them more frequently

–Washing less, letting garments rest and air between use

–Maintaining garments, also repairing

–Using services to intensify use and to extend the use time (e.g. repair, upgrading) (Niinimäki 2013).

Consumption and consumerism lie at the core of Western societies. Consumption is an important function in people's everyday life. Current consumption patterns are strongly connected to industrial manufacturing systems, economic systems and the underlying economic values supporting this unsustainable system. Consumption and purchasing situations often involve a strong emotional experience for consumers. Therefore, we should also create systems that offer other kinds of emotional experiences and satisfaction than from buying new fashion items. This could happen for example through strategically sustainable design that includes services (Niinimäki 2011).

2.10 Designing value

Ehrenfeld (2015) points out that conventional actions in sustainable fashion can at best only slow down the destruction, not save the Earth. According to him, current actions in the industry do not create sustainability, they only reduce unsustainability. He claims that concentrating in, e.g. technical eco-efficiency improvements, CSR or eco-materials only fool the designers, manufacturers and consumers into believing that they are doing all it takes to create sustainability-as-flourishing (*ibid.*, 59). Ehrenfeld (2015) defines sustainability-as-flourishing to mean a normative approach to sustain human beings and ecosystems on the planet forever. In this approach, it should be named what has to sustain. Without naming the elements to be sustained, sustainability is an empty word according to him. To be more influential, sustainable fashion industry has to focus on transforming underlying beliefs and values that currently maintain unsustainable practices in the system level. Design system has to change the beliefs on caring, instead of needing in sustainability-as-flourishing approach. In here, the project 'Local Wisdom' which concentrates on the use phase of clothing and the craft of use is a good example. The project collects user experiences: tending, fixing and satisfying the use of garments in a long term, and it frames the design and use as a single whole and uses the use practices into design thinking, bringing the creative use practices into designers' reality (Flecther 2015). As Fletcher (2008) argues in a more holistic approach to sustainability, the design aesthetic has to be grounded on long-term sustainability and values in it.

Design forms an integral part of our culture and values within. This means that sustainable design should not only take into account environmental values but cultural, social, ethical and economic values should also be included in the sustainable design process (e.g. Tischner and Charter 2001). Designing could change and become a more proactive and participatory process aiming for aesthetic sustainability (Niemelä 2010, 60). To this ambition, it can be added that sustainable design can also be a proactive process that aims to transform consumption patterns towards more sustainable ones, and therefore, it includes future-oriented thinking and future-oriented value (Niinimäki 2011). It is important to understand how the current design, manufacturing and business practices affect unsustainable consumption patterns. Products actually configure consumers' needs and use patterns, and hence, design can be said to be 'practice-oriented', which leads to certain everyday consumption habits (Shove et al. 2007, 134–136). Therefore, designers, manufacturers and companies are responsible not only for the environmental impacts of the fashion industry but also for the unsustainable consumption behaviour of consumers and the increase of waste streams.

Designers can even destroy the value of the product by creating fast-changing trends that quickly outdate the products and render them valueless. The product itself may destroy something valuable, such as a craft skill or local knowledge (Fry 2009), something we already have seen in the area of fashion: manufacturing locations are far from end markets, and local manufacturing skills have diminished or disappeared. At the same time, only few consumers can repair their own garments or have skills to create garments themselves; many do not even own a sewing machine (Niinimäki 2011; Armstrong et al. 2015). Thinking about what we design and produce and why, as well as what consequences our design may bring, forms the basis of ethical thinking and future-oriented value creation.

Future-oriented value creation as an approach guides designers and manufacturers towards far-sighted products and far-sighted business thinking, transforming current practices towards more sustainable societies. Each design and manufacturing decision has to include consideration of the future: how the product will be used, how it will age, how it will be disposed of and what environmental impact the product will have (during manufacturing, use and after use). Furthermore, the business model and its influence on consumption practices are included in this evaluation process (e.g. a slow system versus a fast system). What consequences do the product and its manufacturing have with regard to environmental, societal and cultural values? All these aspects have to be considered before the product is designed or manufactured. And is there the possibility to completely avoid materialistic consumption and do business differently? Radical thinking (e.g. a PSS approach to fulfil consumers' needs in a de-materialistic way) is needed to create a new value creation process in sustainable fashion (Armstrong et al. 2015).

At best, a new kind of sustainable fashion system can create value for people, planet and profit (see Figure 1). For people, the system can create a safe environment through the use of fewer chemicals in production, i.e. safe products for end-users. This system can also promote greater product satisfaction through better quality and longer-lasting products. It can promote well-being and happiness through less materialistic consumption, e.g. by offering experiences through sharing, swapping, lending, updating or do-it-yourself fashion. We can learn to invest in higher quality and more expensive products, use them longer, own less and take good care of our possessions. Furthermore, such a system can mean more social justice and better working conditions for textile factory workers, while through slowing the system, it is possible to increase the end price of the product and invest in better factories. For the environment, through cleaner production and perhaps even less industrial production and waste, the benefits for slower cycles are clear. For business, new models to make profit in an environmentally beneficial way have to be created. New system thinking and these radical business models can afford new business opportunities through, e.g. a service approach, to encourage less materialistic industrial manufacturing (Niinimäki 2013).

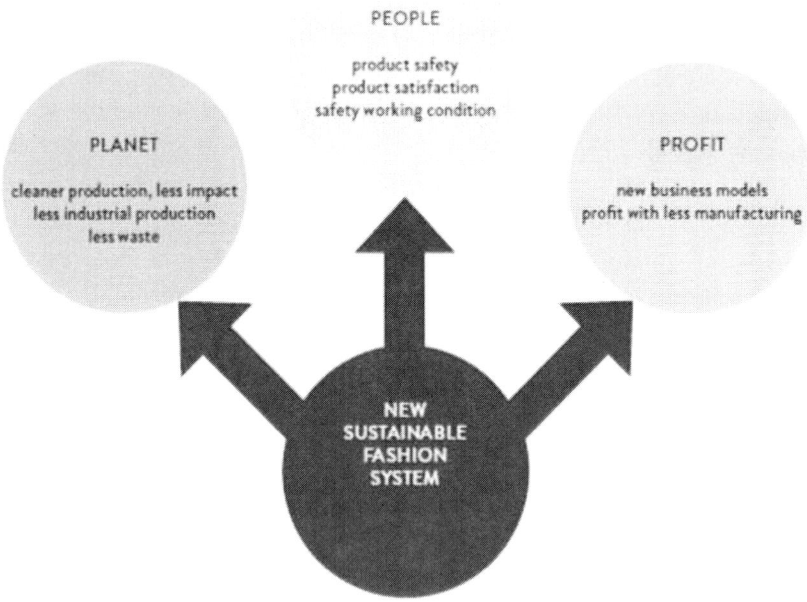

Figure 1. Sustainable fashion system can create value to people, environment and business (Niinimäki, **2013** , 35)nn.

3. CONCLUSIONS

This text provided an overview for ethics in the sustainable fashion field. It was grounded on environmental ethics and value discussion in it. As Ehrenfeld (2015, 61, based on Rittel and Webber 1973) argues 'solutions for wicked problems are not true-or-false but good-or-bad', which grounds the discussion in sustainable design on ethics, moral and values. Moreover, he points out that each 'implemented solution to a wicked problem has consequences'. Sustainable fashion is a wicked problem: there is no simple truth how to make things better or right, but only different approaches to improve the current situation. Values and ethics are fundamental grounds also for sustainable fashion, and they might help to widen the discussion and seeing problem areas more holistically and further to include the ethical dimension more tightly to discussion. Ethics is the ground for making choices in the sustainable design and manufacturing and even while considering the business models and profit-making options. It is worthwhile to investigate the value base in sustainable fashion and further to consider the consequences of our design and industrial manufacturing processes for the environment, human health and for societies.

Based on environmental ethics, first, we have to consider the system level change, what is right for the ecosystem, societies and community as a whole in the context of sustainable development. Secondly, we should consider the human health issues in production and also in the use situation and the use of chemicals in the textile production. Finally, we have to take into account the sustainable use of resources; emphasising renewable and recyclable materials, considering efficient or sufficient principles and further creating a slower and closed-look-system for the fashion industry.

Even though ethical consideration might be a complicated and hard task to do for companies, it can open new business possibilities, while consumers are more than ever interested on environmental values. Products have to offer those values that are important to consumers. If the value offering of the new product is an answer to a consumer's value seeking, there is a business opportunity (Niinimäki 2011). If a consumer's deep inner needs are understood, a new sustainable fashion business can be done in a less materialistic way, e.g. offering services to fulfil the need for appearance change instead of new products or offer emotional experiences without purchasing. Moreover, in the future, in a sustainable economy, producers and consumers begin to be partners in co-producing value in a closed loop industrial system where all materials, even old clothes, are valuable raw material for the industrial system.

The values behind the product and company create the foundation upon which a good design is built: a design that is grounded on ethical consideration, which has a low environmental impact and which is also easy to fall in love with. This is the way to do sustainable fashion in the future.

COMPETING INTERESTS

The author declares that she has no competing interest.

REFERENCES

1. Allwood, JM, Laursen, SEM, de Rodriguez, C, & Bocken, NMP. (2006). *Well dressed?* Cambridge, UK: University of Cambridge Institute of Manufacturing.

2. Armstrong, C., Niinimäki, K., Kujala, S., Karell, E. & Lang, C. (2015). Sustainable Product-Service Systems for Clothing: Exploring Consumer Perceptions of Consumption Alternatives in Finland. Journal of Cleaner Production (in press).

3. Belz, FM, & Peattie, K. (2010). *Sustainability marketing. A global perspective*. Chichester, UK: John Wiley & Son.

4. Belz, F, & Peattie, K. (2011). *Sustainability marketing: a global perspective* (3rd ed.). West Sussex, UK: John Wiley & Sons.

5. Black, S. (2011). *Eco-chic: the fashion paradox*. London, UK: Black Dog Publishing.

6. Brezet, H, & Van Hemel, C. (1997). *Ecodesign: a promising approach to sustainable production and consumption*. Paris: UNEP.

7. Clegg, B. (2011). *Sustainable business: financial times briefings*. Harlow, UK: Pearson education.

8. Cooper, T. (2005). Slower consumption: reflections on products' life spans and the 'throwaway society'.*Journal of Industrial Ecology, 9*(1–2), 51–67.

9. Ehrenfeld, JR. (2015). The real challenge of sustainability. In K Fletcher & M Tham (Eds.), *Routledge handbook of sustainability and fashion* (pp. 57–63). New York, USA: Routledge.

10. Flecther, K. (2015). Other fashion system. In K Fletcher & M Tham (Eds.), *Routledge handbook of sustainability and fashion* (pp. 15–24). New York, USA: Routledge.

11. Fletcher, K. (2008). *Sustainable fashion & textiles: design journeys.* London: Earthscan.

12. Fletcher, K, & Grose, L. (2012). *Fashion & sustainability: design for change.* London, UK: Laurence King Publishing.

13. Fry, T. (2009). *Design futuring: sustainability, ethics and new practice.* Oxford: Berg.

14. Gibbins, J, & Reimar. (1999). *The politics of postmodernity.* London, UK: An introduction to contemporary politics and culture.

15. Goworek, H, Hiller, A, Fisher, T, Cooper, T, & Woodward, S. (2013). Consumers' attitudes towards sustainable fashion. In M Gardetti & A Torres (Eds.), *Sustainability in fashion and textiles* (pp. 377–392). Sheffield, UK: Greenleaf Publishing.

16. Jackson, T, & Shaw, D. (2009). *Mastering Fashion Marketing.* New York, USA: Palgrave Macmillan.

17. Kemp, R. (2008). Transition management for sustainable consumption and production. In A Tukker, M Charter, C Vezzoli, E Stø, & MM Andersen (Eds.), *System innovation for sustainability: perspectives on radical changes to sustainable consumption and production* (pp. 369–390). Sheffield, UK: Greenleaf.

18. Kruger, H, Himmestrup Dahl, E, Hjort, T, & Planthinn, D. (2012). *Guide lines II: a handbook on sustainability in fashion.* Copenhagen: Sustainable Solution Design Association, SSDA.

19. Leiserowitz, AA, Kates, RW, & Parris, TM. (2006). Sustainability values, attitudes, and behaviors: a review of multinational and global trends. *Annual Review, Environment and Resources, 31*, 413–

444. http://environment.yale.edu/climate-communication/files/AttitudesSustainDevelop.pdf (Accessed 1 March 2011).

20. Light, A, & Rolston, H, III. (2003). Introduction: ethics and environmental ethics. In A Light & H Rolston III (Eds.),*Environmental ethics. An anthology* (pp. 1–11). Oxford, UK: Blackwell Publishing.

21. Lowitt, E. (2011). *The future of value: how sustainability creates value through competitive differentiation.* San Francisco: Jossey-Bass.

22. Manzini, E. (1994). Design, environment and social quality: from "existenzminimum" to "quality maximum".*Design Issues, 10*(1), 37–43.

23. Markides, CC. (1997). *Strategic innovation. Sloan Management Review. Spring, 1997*, 9–23.

24. Niemelä, M. (2010). *Kestävää muotoilua mallintamassa. Tulkitseva käsitetutkimus taideteollisen muotoilun näkökulmasta. [modeling sustainable design. Interpretative concept research from the viewpoint of craft design]. doctoral dissertation. Helsinki*. Finland: University of Art and Design Helsinki.

25. Niinimäki, K, & Aakko, M. (2014). *Creative control in sustainable fashion. 19th ADMC, academic design management conference, design management in the era of disruption. 2–4 September 2014* (pp. 583–600). London, UK: UAL.

26. Niinimäki, K. (2011). *From disposable to sustainable: the complex interplay between design and consumption of textiles and clothing. Doctoral dissertation.* Helsinki: Aalto University. https://aaltodoc.aalto.fi/handle/123456789/13770 (Accessed 1 October 2014).

27. Niinimäki, K (Ed.). (2013). *Sustainable fashion: new approaches.* Helsinki: Aalto ARTS Books. https://aaltodoc.aalto.fi/handle/123456789/13769(Accessed 1 October 2014).

28. Niinimäki, K. (2014). Sustainable consumer satisfaction in the context of clothing. In C Vezzoli, C Kohtala, & A Srinivasan (Eds.), *Product-service system design for sustainability* (pp. 218–237). Sheffield, UK: Greenleaf.

29. Ottoman, JA. (2011). *The new rules of green marketing: strategies, tools, and inspiration for sustainable branding*. Sheffield, UK: Greenleaf.

30. Paehike, RC. (2000). Environmental values and public policy. In NJ Vig & ME Kraft (Eds.), *Environmental Policy*(4th ed., pp. 77–97). Washington, USA: CQ Press.

31. Palmer, C. (2003). An overview of environmental ethics. In A Light & H Rolston III (Eds.), *Environmental ethics. An Anthology* (pp. 15–37). Oxford, UK: Blackwell Publishing.

32. Papanek, V. (1995). *The green imperative. Ecology and ethics in design and architecture*. London: Thames and Hudson.

33. Park, P, & Tahara, K. (2008). Quantifying producer and consumer-based eco-efficiencies for the identification of key ecodesign issues. *Journal of Cleaner Production, 16*, 95–104.

34. Porter, ME, & Kramer, MR. (2006). Strategy and society; the link between competitive advantage and corporate social responsibility. *Harvard business review, 84*(12), 78–92. 163.

35. Scherer, D. (2003). The ethics of sustainable resources. In A Light & H Rolston III (Eds.), *Environmental ethics. An Anthology* (pp. 334–358). Oxford, UK: Blackwell Publishing.

36. Shove, E, Watson, M, Hand, M, & Ingram, J. (2007). *The design of everyday life*. Oxford: Berg.

37. Tischner, U, & Charter, M. (2001). Sustainable product design. In M Charter & U Tischner (Eds.), *Sustainable solutions: developing*

products and services for the future (pp. 118–138). Sheffield: Greenleaf.

38. Tojo, N, Kogg, B, Kiørboe, N, Kjær, B, & Aalto, K. (2012). *Prevention of textile waste: material flows of textile in three Nordic countries and suggestions on policy instruments.* Copenhagen: Nordic Council of Ministers.

39. Tukker, A, Emmert, S, Charter, M, Vezzoli, C, Stø, E, Andersen, MM, et al. (2008). Fostering change to sustainable consumption and production: an evidence based view. *Journal of Cleaner Production, 16,* 1218–1225.

40. van Dyk, L. (2008). Developing strategies for a typology of sustainable fashion design. In J Hethorn & C Ulasewicz (Eds.), *Sustainable fashion. Why now?* (pp. 233–263). New York: Fairchild Books.

41. Verganti, R. (2009). *Design driven innovation: changing the rules of competition by radically innovating what things mean.* Boston, USA: Harvard Business Press.

42. Vezzoli, C. (2007). *System design for sustainability. Theory, methods and tools for a sustainable "satisfaction-system" design.* Milan: Maggioli Editore.

43. Walker, S. (2007). *Sustainable by design. Exploration in theory and practice.* London, UK: Earthscan.

4

ENVIRONMENTAL SUSTAINABILITY EVALUATION OF APPAREL PRODUCT: A CASE STUDY ON KNITTED T-SHIRT

Md. Mazedul Islam[1] and Md. Mashiur Rahman Khan[2]

[1]Department of Textile Engineering, Dhaka University of Engineering & Technology, Gazipur 1700, Bangladesh

[2]Department of Apparel Manufacturing Engineering, Bangladesh University of Textiles, Tejgaon, Dhaka 1208, Bangladesh

ABSTRACT

The work reported in this paper presents the assessment of environmental performances of selected branded apparel T-Shirt products made by Bangladesh. The study is based on a standard evaluation tool named Higg Index which is basically used widely to measure the environmental sustainability of various apparel products. Higg Index is an internal self-assessment tool created by the outdoor apparel industry and Nike's apparel environmental design tool which aims to aggregate information on the environmental performance of products. The Index considers performance across the full life-cycle of a product, including impacts from "input materials, manufacturing, packaging, transportation, use, and end-of-life." Selected apparel branded T-Shirt products from S. Oliver, BUTex-Innovation, PUMA, Esprit, Aarong, and Yellow were taken into consideration. The results indicate that newly developed ecofriendly T-shirt and foreign branded products named S. Oliver, PUMA, and Esprit gained higher score but local branded product like Aarong and Yellow gained lower score in terms of environmental sustainability based on Higg Index assessment tool. Moreover, many weaknesses and opportunities for improvement of both local and foreign branded T-Shirt products have been identified and suggested which would eventually lead the fashion designer, apparel manufacturer, stakeholder, and consumer towards greener apparel products.

1. INTRODUCTION

Sustainability is the major concern in the age of modern world. For textile and apparel sector, this has been a burning issue for many related concerned bodies. Over the past few years, increasing awareness of the environmental and social concerns surrounding the fashion industry have led to a rise in the implementation of sustainability initiatives. There has been a growing concern over apparel brands in improving their environmental impact and the social responsibility throughout their supply chains. Environmental sustainability refers to the ability of something to continue without upsetting earth's ecological balance. Sustainable apparel products can be defined as a part of the growing design philosophy and trend of sustainability, the goal of which is to create a system which can be supported indefinitely in terms of environmentalism and social responsibility [1]. Environmental sustainability in business refers to longevity, but in terms of which natural resources the production process might draw upon, how resources are used and replenished, the overall impact of the final production on the environment, and where the product ends up following its disposal. The textile and apparel products impact the environment at every point along the product's lifecycle [1]. However, rapid growth in the industrial sector is playing a vital role in the economy of Bangladesh. Despite the growing necessity of environmental practices in the apparel industry, little research has explored methods and standard tools to guide apparel designers, manufacturers, merchandisers, and consumers in adapting environmental issues in their work. Many designers do not recognize how their designs impact the environment. The role of the designer must expand dramatically in light of current environmental concerns with the contamination and destruction of the ecosystem [2]. Today's designer, manufacturer, merchandiser, and consumer must be well informed on a vast array of topics, from methods of production to governmental regulations to life cycle analysis. They must account for the environmental impacts of the materials uses, the resulting waste from the forms they chooses, how products are produced and packaged, where they will be made and then sold causing energy use for transportation, and how consumers may use and dispose of the product [3]. Hence, there is a growing need to measure the environmental sustainability of apparel products by standard methods, tools, and technique to minimize the environmental impact. The Higg Index by sustainable apparel coalition is such a standard tool by which the environmental sustainability of apparel products could be measured. This study attempts to identify and analyse a number of these issues related to apparel product sustainability taking several local and foreign branded T-shirts into consideration as case study.

2. LITERATURE REVIEW

2.1. Sustainable Apparel Product

Sustainability is defined as the design of human and industrial systems to ensure that humankind's use of natural resources and cycles do not lead to diminished quality of life due either to losses in future economic opportunities or to adverse impacts on social conditions, human health, and the environment [3]. To achieve sustainable development, designers need to be aware of environmental impacts and incorporate environmental awareness into the design [2]. Product sustainability is the easiest aspect to alter for an apparel brand, as this is where a company has the most and direct control through design and product development [2]. Transforming product sustainability may be achieved via various aspects such as fiber/textile selection, processing methods, use behaviors, and reuse/recycle strategies. Fiber/textile selection is often the first step that designers and product developers will take in reducing the environmental impact of a garment. Environmentally preferred fibres/textiles can significantly reduce the environmental impact and increase the resourcefulness of an apparel product throughout the garments life cycle without change to design practice or product development processes [3].

2.2. Environmental Impact of Textiles Apparel Products

The apparel industry has many negative environmental and social impacts that are complex and occur at different stages of the apparel life cycle. The development of fast fashion has amplified the impacts due to the increased volume of apparel produced and sold at low prices [4]. Apparel has a long and complicated production chain consisting of many phases including resource production and extraction, fibre and yarn manufacturing, textile manufacturing, apparel assembly, packaging, transportation and distribution, consumer use, recycling, and ultimate disposal [5]. The environmental impacts of apparel are varied across the phases, difficult to assess for individual garments, and subject to type of raw material used, dyeing, and laundering. It was not until the 1990s that there was a greater awareness as to the severity of the negative environmental impacts of apparel production [3]. The major environmental impacts associated with the production and use of apparel throughout its life cycle includes wastewater emissions from dyeing, finishing, and washing processes, increase in pollution, solid waste production, and significant depletion of resources from consumption of water, fossil fuels, and raw materials [2, 6]. Energy is used for laundering, transportation, operations of machines for various processes, production of primary materials especially man-made fibres such as polyester (a petroleum based product) and yarn manufacturing of natural fibres such as cotton. Conventional cotton production has high water consumption and employs the use of toxic chemicals that may harm human health and the environment [1, 7]. Chemicals are also released in wastewater from processes such as pretreatments, dyeing, finishing, and laundry. These chemicals are disruptive to both the environment and aquatic-based life [8]. Solid waste is produced during natural fibre yarn, textile, and apparel manufacturing and disposal of apparel products at the end of their life. There are significant issues with clothing waste as the majority of clothing and textile waste ends up in landfills as opposed to being recycled or reused [2].

2.3. Sustainable Apparel Coalition and Higg Index Tool

The SAC is a multistakeholder engagement, formed in 2011, by a group of global apparel and footwear companies and nonprofit organizations representing nearly one third of the global market share for apparel and footwear. The SAC seeks to build a common approach for measuring and evaluating apparel and footwear product sustainability performance [9]. It aims to develop common measurements and a common environmental understanding of products' impacts across the industry by building on the Outdoor Industry Association (OIA)'s Eco Index and Nike's Environmental Design tool [10]. The Eco Index is a standardized tool for measuring the environmental impacts of outdoor products such as boots, clothing, and tents and evaluates the impacts in six key areas of a product's life cycle: materials, packaging, product manufacturing, and assembly, transport and distribution, use of service, and end of life [9]. Nike's Environmental Design tool measures the environmental impacts of apparel. Measuring performance of apparel products will spotlight priorities for action and opportunities for technological innovation [9]. The Higg index 1.0 is an "indicator based tool for apparel that enables companies to evaluate material types, products, facilities, and processes based on a range of environmental and product design choices" [9]. The scope for the desired outcomes of the SAC Higg Index 1.0 tool includes improvements to reduce water use and improve quality, reduce energy and emissions, minimizing waste, reduce chemicals and toxicity, and increase transparency for social and labour issues [6]. The first version of the Higg Index 1.0 tool which was released in June 2012 was based on life cycle thinking and is publicly available for any organization. The Index tool was developed to measure the environmental impacts of apparel products. The tool evaluates material type, products, and facilities and processes. By utilizing practice based, qualitative binary yes/no questions, assessments can be made as to the sustainability performance of product and drive behavior for improvement. The Higg Index RDM-Beta is a prototype aims at educating and providing quick directional guidance to apparel designers during the product creation process about the potential environmental impacts of their design solutions [11]. It is to engage designers in the Higg Index's produce life cycle thinking and how we directionally assess materials sustainability through the MSI. By helping organizations standardize measurement and evaluation of environmental performance, the tool creates a starting point for engagement, education, and collaboration among stakeholders [12]. Through self-assessment, organizations can better understand the environmental impacts that occur throughout the life cycle and the effect of design choices such as material type. However, there are many apparel brands belonging to the SAC such as H&M, Gap Inc., Nike, adidas, Puma, Patagonia, Mountain Equipment Co-op (MEC), Levi's, Hanesbrand, Marks & Spencer, Esprit, Columbia, Timberland and Loomstate, and S. Oliver. These brands are widely using this Higg Index tool to measure environmental performance of their apparel products towards better sustainability [11].

3. MATERIALS AND METHODOLOGY

To carry out this study, selected knitted men's T-shirt products of 100% cotton single jersey fabric were taken into consideration as case study. Both foreign and local branded knitted men's "M" sized short sleeve T-shirt products, namely, S. Oliver, BUTex Innovation, PUMA, Levi's, Aarong, and Yellow, were selected on random basis to measure their environmental performance. The PUMA, S. Oliver, and Esprit branded T-shirts were collected from various organizations, namely, Viyellatex group Ltd. And Yellow and Aarong T-shirts were collected from Beximco fashion wear and Aarong and BUTex Innovation T-shirts were developed by the researcher as ecofriendly. Higg Index provides a value that represents how environmentally responsible an apparel item is in terms of sustainability. For this study, the tool was used to obtain product scores for the above mentioned T-Shirt. This score was obtained by answering questions relating to the materials, packaging, manufacturing, product care, and repair service, and product end of life.

The data used during this process includes information readily available to the general consumers, manufacturers and merchandisers including hangtags, care labels, and online product descriptions and throughout the LCA phase of the developed T-Shirt. The product Tec-pack details information was the important sources for various Higg Index required information towards sustainability measurement. Higg Index 1.0 requires that all materials present be entered into a materials input table. This includes a description of all materials used in addition to information obtained from follow-up questions. By utilizing practice based, qualitative binary yes/no questions, assessments can be made as to the sustainability performance of product and drive behavior for improvement. Higg Index 1.0 is a Microsoft Office Excel and wed based tool that has predetermined standard values based on global survey for specific materials that are used in determining a score. The higher the Higg Index score, the better the environmental performance of an apparel products.

4. RESULTS AND DISCUSSIONS

In this study, all the five selected foreign and local branded knitted T-shirts alone with new developed one were assessed using the Higg Index 1.0 to obtain a value that represents a product's environmental impact. A higher score implies a more sustainable product. Upon completion of the excel and web based Higg Index tool, the BUTex-Innovation, S. Oliver, PUMA, Esprit, Yellow, and Aarong T-shirts received total score of 75, 57, 79, 53, 47, and 36, respectively, on a scale of 1–100. The superior score of the BUTex Innovation and foreign banded product especially PUMA T-shirt was due to the organic material content, ecofriendly raw materials and process used product end of life options and confirmed traceability. On the other hand the local branded T-shirt gained comparatively lower score in most of the areas due not using ecofriendly raw materials, manufacturing processes, high impact finishes, improper or insufficient care and repair instructions, traceability issues, and so forth. This also confirms that by increasing the transparency of a product and offering more information about a product's development, the more this product will represent ecofriendliness and will be superior in terms of sustainability.

4.1. Materials Selection

The materials selected for the knitted T-shirt for both foreign and local branded product were entered into the material inputs table in the Higg Index 1.0. tool. The values obtained by various products are shown in Table 1. The developed men's T-shirt and foreign brands PUMA, S. Oliver, and Esprit obtained higher score. This was due to the ecofriendly raw materials used, that is, BUTex Innovation and PUMA t-shirt included 100% organic cotton fibre. During processing of fabric, BUTex Innovation T-Shirt, PUMA, and other foreign branded T-shirts used ecofriendly and certified and environmentally low impacted dyes chemicals. But in case of local branded Yellow and Aarong T-shirts Higg Index score was lower due to environmentally hazardous conventional raw materials, processing, dyes used, no third party verification, and so forth. For better environmental sustainability these issues must be addressed for local branded apparel products.

4.2. Manufacturing Stages

The manufacturing stages are obviously an important area for better product sustainability. Some major issues have great impact on product environmental sustainability as per Higg Index tool. The developed men's T-shirt, PUMA, and foreign branded T-shirt used low impact garment finish, that is, process combination enzyme wash or basic rinse that scored higher. But local branded Aarong and Yellow T-shirt used high impact garment finish, namely, traditional bleach, acid chemical wash, and so forth and scored lower. Besides, marker efficiency has great impact on product sustainability. The higher the marker efficiency, the better the product sustainability and lower marker efficiency indicate its direct impact on supply chain including more wastages of material like fabric, dyes chemicals, process loss, land fill, and so forth. As developed T-shirt, PUMA, S. Oliver, and Esprit brand had more than

92% marker efficiency, they scored better. But Aarong and Yelllow brands had marker efficiency of 78% and 84% and scored lower in the Higg Index tool. Similarly, a single colored, screen printed logo with small area received a higher score of for foreign branded T-shirt and these issues scored lower for local branded T-shirt. The total points for manufacturing stages are shown in Figure 1.

4.3. Packaging

Product packaging was included in this assessment. Conventional and environmentally harmful materials were used for local branded T-shirt especially in case of Aarong, which did not provide any score points rather than zero. But PUMA and other T-shirt used only sustainable, third party verified and reduced packaging materials and obtained better score. The packaging scores are shown in Table 1.

4.4. Product Care and Repair Services

For PUMA, new developed men's T-Shirt, S. Oliver, and Esprit T-shirts, best class to low impact details care instructions were provided due to the required care of the chosen materials that scored higher, that is, 18 for PUMA. But for Yellow and Aarong T-shirts, there were no in details care instructions that leaded to lower score in Higg Index, that is, only 4. Proper care instructions should be provided to the product label as most of the water and energy is used by the customer during the care and repair service stages.

Table 1: Higg Index evaluation score for T-shirts products

Scoring summary stages	Section score each out of 100						Section weight %	Score obtained					
	BUTex-Innovation	S. Oliver	PUMA	Esprit	Yellow	Aarong		BUTex-Innovation	S. Oliver	PUMA	Esprit	Yellow	Aarong
Materials selection	100	71	100	71	66	57	30%	30	21	30	21	20	17
Manufacturing (i) Marker efficiency (ii) Garment finish (iii) Applied graphics and logo (iv) Screen printing size and color	53	55	53	53	40	24	25%	13	14	13	13	10	6
Packaging	90	90	90	50	40	0	10%	9	9	9	5	4	0
Product use and care service	90	20	90	20	20	20	20%	14	4	18	4	4	4
End of life	60	60	60	60	60	60	15%	9	9	9	9	9	9
Total							100%	75	57	79	53	47	36

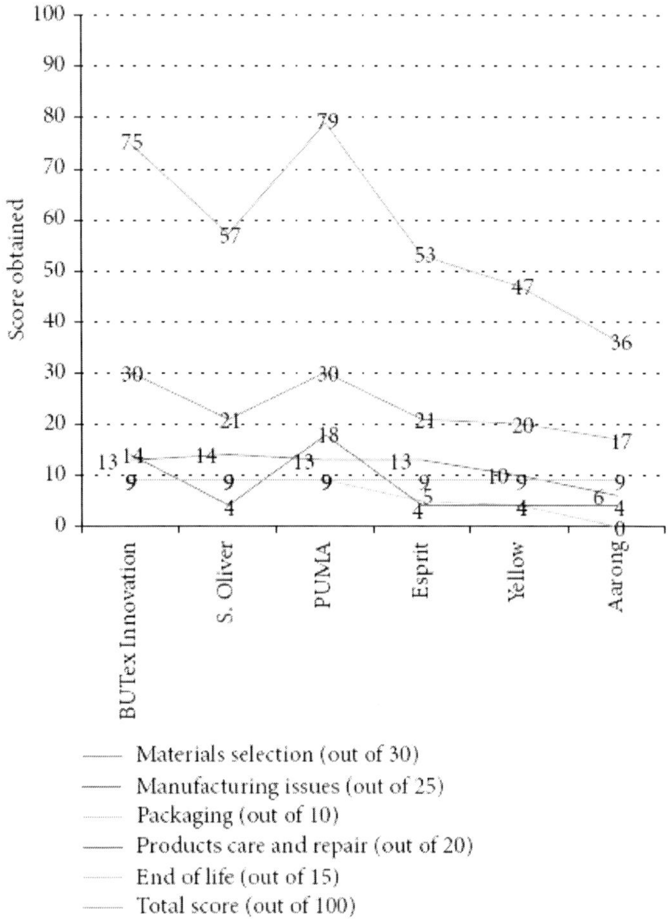

Figure 1: Higg Index assessment score for various branded T-Shirt.

4.5. End of Life

All T-shirt products received 60 points and a base score 9 for product end of life. The PUMA and BUTex Innovation product is made from 100% organic cotton and other branded T-shirt made from conventional cotton. These individual material types can be accurately identified by end of use facilities or processes. The material types are recycled via existing infrastructure and processes or materials are used in closed-loop recycling processes were not known. In response to the questionnaires regarding end of life of products as per assessment Higg Index tools, the all products responded in the same way to the questionnaires. So the EOL stages obtained a base score, that is, 60%, and no additional points were given as shown in Table 1.

5. TECHNICAL ISSUES FOR APPAREL PRODUCTS SUSTAINABILITY IMPROVEMENT

The researcher has identified many technical requirements issues for apparel products sustainability improvement based on Higg Index assessment tool.

Sourcing Fair Production and Practice Transparency. Sourcing fair production can be performed by selecting manufacturers that are known to operate in a socially responsible manner. For the technical requirement of practicing transparency, the researcher suggests a hangtag including maximum in details information to the products that includes the name and contact information of a hypothetical manufacturer. This would allow a customer to verify socially responsible methods were practiced when the product was produced.

Sourcing Local Production. Local production is a valuable asset to improve product sustainability. This will also improve the social and economic sustainability of a product.

Consider Waste Reducing Design and Clean Production Adjustments. To reduce waste, the researcher suggested using ecofriendly dyeing method like reduced water dyeing or waterless dyeing method for the fabric processing. An on-product label with instructions to recycle, reuse of the product, thus reducing waste, should be added to the label. Clean production can be achieved by selecting manufacturers that make a commitment to the use of alternative or renewable energies.

Consider Organic and Natural Fibers. Natural fibers are often viewed as the most environmentally responsible as they come from renewable sources and use less energy and chemicals to produce than manufactured fibers. Ecofriendly fibers, production methods that demand fewer natural resources and use nature to their benefit should be used.

Source of CMT Production. In order to replicate the possible effects of sourcing CMT production, the researcher decided to identify only one manufacturing facility instead of two or more in the Higg Index 1.0 to determine if fewer facilities results in a better score.

Consider Natural and Low Impact Dyes and Chemicals. The designers, manufacturers, and merchandiser should introduce natural and environmentally low impact dyes in materials processing to better product sustainability.

6. CONCLUSION

The textile and apparel industries are notorious for their excessive contribution of waste and pollution to our environment. The unsustainable practices trends of poorly designing, raw materials selection, processing, manufacturing, and disposing of apparel product is a major contributing factor to this problem. The purpose of this study was to evaluate the environmental sustainability of various branded T-shirts and find out various sustainable issues for greener apparel products. Based on Higg Index environmental assessment standard tool, the findings of this study could be helpful for many organizations and customers who attempt to produce and consume sustainable textiles and clothing products. Both the local and foreign branded products should be produced by strictly maintaining sustainability issues. Hence, this study is a little approach towards better sustainable practices and eventually would drive the textile and clothing designer, manufacturers, merchandisers, and consumers to introduce various sustainable issues in their works. Hence, growing awareness and leading them towards sustainable practices in the textile and clothing business for greener world.

CONFLICT OF INTERESTS

The authors declare that there is no conflict of interests regarding the publication of this paper.

REFERENCES

1. J. M. Allwood, S. E. Laursen, C. M. de Rodriguez, and N. M. P. Bocken, Well Dressed? The Present and Future Sustainability of Clothing and Textiles in the United Kingdom, Institute of Manufacturing, University of Cambridge, Cambridge, UK, 2006.

2. K. Fletcher, Sustainable Fashion & Textiles, Earthscan, London, UK, 2008.

3. C. A. Rusinko, "Green manufacturing: An evaluation of environmentally sustainable manufacturing practices and their impact on competitive outcomes," IEEE Transactions on Engineering Management, vol. 54, no. 3, pp. 445–454, 2007.

4. J. K. Steinberger, D. Friot, O. Jolliet, and S. Erkman, "A spatially explicit life cycle inventory of the global textile chain," The International Journal of Life Cycle Assessment, vol. 14, no. 5, pp. 443–455, 2009.

5. V. D. Lewis and C. Chen, "The life of a piece of cloth: developing garments into a sustainable service system," International Journal of Environmental, Cultural, Economic and Social Sustainability, vol. 2, no. 1, pp. 197–207, 2006.

6. H. Goworek, "Social and environmental sustainability in the clothing industry: a case study of a fair trade retailer," Social Responsibility Journal, vol. 7, no. 1, pp. 74–86, 2011.

7. J. Speer, "Organic cotton: where, why and how. Appare," Journal of Sustainable Textiles Materials, vol. 46, no. 9, pp. 29–34, 2005.

8. S. E. Laursen, J. Hansen, H. H. Knudsen, H. Wenzel, H. F. Larsen, and F. M. Kristensen, EDIPTEX: Environmental Assessment of Textiles, Danish Environmental Protection Agency, Copenhagen, Denmark, 2007.

9. The higg index, "Sustainable Apparel Coalition," 2012, http://www.apparelcoalition.org/higgindex.

10. K. Niinimäki and L. Hassi, "Emerging design strategies in sustainable production and consumption of textiles and clothing," Journal of Cleaner Production, vol. 19, no. 16, pp. 1876–1883, 2011.

11. Reuben, The Higg Index for Sustainable Apparel Environmental Performance Index, 2013,http://epi.yale.edu/indicators/indicator%E2%80%90case%E2%80%90studies/reports/higg%E2%80%90index.

12. A. Adams and G. R. Frost, "Integrating sustainability reporting into management practices,"Accounting Forum, vol. 32, no. 4, pp. 288–302, 2008.

5

THE SUSTAINABLE FUTURE OF THE SCOTTISH TEXTILES SECTOR: CHALLENGES AND OPPORTUNITIES OF INTRODUCING A CIRCULAR ECONOMY MODEL

Lynn Wilson

Zero Waste Scotland, Ground Floor, Moray House, Forthside Way, Stirling FK8 1QZ, UK

ABSTRACT

Case description

Zero Waste Scotland introduced the concept of the circular economy to the Scottish textiles sector at events throughout 2013 to 2014. In April 2014, it commissioned research by independent consultants to examine the academic and industrial textile landscapes in Scotland, including developments in technical textiles and research into innovation in textile design and examples of circular economy models.

The research identified a number of initiatives, including projects producing an alternative to denim and one developing cavity wall insulation from processed natural fibres. It made recommendations to Zero Waste Scotland about shaping the future landscape of textile innovation in Scotland and also offered examples of the circular economy from Scandinavia that might be applicable.

Discussion and evaluation

The implementation of the circular economy into Scotland's textile sector, underpinned by the aforementioned initiatives, can learn too from its textile and fibre heritage. Zero Waste Scotland is implementing an action plan which will offer support to the textile industry exploring 'closed loop' manufacturing, as well as funds for fashion designers to explore concepts such as zero-waste pattern design, luxury apparel from alternative textiles such as recycled polyethylene terephthalate and natural fibres such as nettle. A master class skills programme, delivered by leading UK and international experts, will bring together industry, academia, and higher education professionals to engage in learning and information exchanges about the circular economy. This paper presents Zero Waste Scotland's role and the research findings.

KEYWORDS

Circular economy, Scottish textiles sector, Closed-loop manufacturing

1. BACKGROUND

Zero Waste Scotland (ZWS) was initially a programme of the Waste Resource Action Programme UK (WRAP UK). On 1 July 2014, ZWS was formally launched as an independent Scottish not-for-profit company.

On 11 February 2014, WRAP UK launched the Sustainable Clothing Action Plan, SCAP 2020 Commitment. Sixty-five signatories and supporters of the SCAP 2020 Commitment have pledged to reduce the carbon, water and waste footprints of clothing they supply or receive in the UK by 15% by 2020. Six of these companies are from Scotland.

ZWS supports SCAP 2020 and WRAP UK's Love Your Clothes campaign 2014 to 2016. In 2014, its key partners were John Lewis and Young Scot, the Scottish Youth information portal for 11 to 26 year olds. Both organisations supported the ZWS organised 'Spring Clean Your Wardrobe Weekend'.

ZWS' textiles portfolio and work programme for 2014 to 2016 is managed by its Circular Economy Team, which also covers Oil and Gas Decommissioning and Circular Economy Business Models. It identified various economic reports available from Scottish Enterprise (SE) and Scottish Development International (SDI) on the textiles industry in Scotland[a]. However, there are few reports related to circular economy innovation in textiles in Scotland and evidence of closed-loop manufacturing.

In April 2014, ZWS commissioned independent research consultants to undertake an examination of the academic and industrial textile innovation landscapes in Scotland, including developments in technical textiles and research into innovation in textile design.

The aim of the commissioned research was to contribute to informing the direction and development of the 2014 to 2016 work programme delivered by ZWS' Circular Economy Team with the textiles sector in Scotland. The report, submitted to ZWS August 2014, examined the circular economy in relation to the textiles sector in Scotland.

The circular economy is, of course, more than just another name for 'reuse'. The Ellen McArthur Foundation says of it:

The circular economy is a generic term for an economy that is regenerative by design. Materials flows are of two types, biological materials, designed to reenter the biosphere, and technical materials, designed to circulate with minimal loss of quality, in turn entraining the shift towards an economy ultimately powered by renewable energy. [b]

The report submitted to ZWS found that there are pockets of innovation in the textiles sector in Scotland that are contributing to a circular economy; however, mainly, the driver is a commercial business goal rather than corporate social responsibility. This may reflect the relative weakness of the Scottish textile industry over the last decade and the lack of an overarching strategic vision for the industry as well as one that is drawing on, and support of, academic research.

Specifically, the report noted variations within the Scottish textiles industry's four sub-sectors: design, heritage/aesthetic, manufacturing, and technical textiles.

It highlighted the following:

1) Examples of innovation or resource-efficient practice already happening in Scotland, such as (a) in its strong technical textiles - where opportunities may exist for closed-loop manufacturing and (b) computer-aided design.

2) Scotland should strengthen what it already does well and bolster its sustainability activities by linking to provenance, traceability, durability and quality.

3) But it also said that 'no true examples of circular economy business models in Scotland were identified'. To that end, it recommended that the search should be broadened to Europe.

4) It is hoped the creation of the Textiles Innovation Council will 'be instrumental in driving change towards a circular economy'[c].

The report, together with ZWS' analysis of its work with the textiles sector up to 2014, highlighted the challenges, opportunities and innovations that can make the Scottish textile industry more materially circular. For example, ZWS has been investigating innovative ways to protect and conserve resources and develop closed-loop manufacturing and high-value products and support services which contribute to a sustainable circular economy for the textiles sector.

In a broader context, this work is part of ZWS' support of the Scottish Government's Zero Waste Plan which has set ambitious targets to achieve 70% recycling and maximum 5% to landfill by 2025 for *all* Scotland's waste. 'Safeguarding Scotland's Resources', launched on 9 June 2013, is the Scottish Government's programme to reduce waste and create a more productive and circular economy[d] with one clear benefit: 'There is a potential saving of £2.9 billion through straightforward resource efficiency that this programme will help tap.' The Scottish Government has sealed its commitment by being the first government to join the Ellen MacArthur Foundation CE 100[e].

2. CASE DESCRIPTION

The research submitted to ZWS has informed its work programme for 2014 to 2016 (published internally August 2014) in terms of themes, policy, master classes, training events, campaigns and engagement with both industry and academia. The key objectives of the programme include the following:

To scope the potential for fibre processing in Scotland and early enquiry around establishing a sustainability standard for Scotland

To showcase and pilot further industry models of a circular economy (CE) and resource-efficient good practice

To seek greater engagement of industry *with* academia in terms of CE practice and theory

ZWS' textiles portfolio includes the following:

- The Scottish textiles sector
- Bulky textile waste - carpets and mattresses, public sector uniform procurement
- Commercial and consumer behaviour change

ZWS feels that the textiles sector has a significant part to play since it is the seventh most important contributor to the Scottish economy, with an annual turnover of £956 million. Exports of Scottish textile products are valued at £375 million. There are over 570 companies directly employing around 9,000 people across Scotland, with 55% of Scottish textile companies having fewer than 10 employees.

In areas such as the Scottish Borders, 50% of employment is within the textile industry. Recent investments from international global brands such as Chanel[f] demonstrate the high value and lucrative position of the Scottish textiles sector. Chanel announced at the beginning of 2014 that they would create 100 new jobs over the next 3 years at their plant in Hawick.

Global economic trends too have had a positive impact on the luxury export side of the Scottish textiles sector. In October 2013, the sector reported a level of growth 12% above the 2020 target and as a result, it has revised its targets to achieve between £1.2 and £1.5 billion in turnover growth by 2020. These revised figures recognise the significant gains to be made in international trade, and a 50% increase in exports is predicted by 2017[g]. Sixty-four percent of textile production in Scotland is for luxury export goods[b].

This increased productivity could contribute to the global waste and environmental impacts that textiles and apparel production create but it can be reduced through the design and manufacturing of products that follow circular economy guidelines or the creation of products with a low carbon impact. Scotland is already developing key products and customers in this market, such as hand-knitted apparel from Fair Isle and the Harris Tweed Authority, which reported an increase of 25% in sales in 2013.

The carbon and water impacts of the textiles supply chain and production are also significant, particularly within the luxury Scottish market, which relies heavily on importing all raw materials such as cashmere from Mongolia and China. Scotland imported £1.4 million of the highest-quality cashmere from Mongolia in 2012/2013, and the political importance of this was highlighted by a visit from a Mongolian government delegation to Scotland hosted by the Scottish Government in 2013[i].

The implementation of the circular economy within the textiles sector not only relies on the aforementioned initiatives and ZWS' work programmes but could also benefit substantively from reflecting on how Scotland's four sub-sectors that the ZWS-commissioned report identified, mentioned above, could be integrated into such an approach.

ZWS introduced the concept of the circular economy to the Scottish textiles sector through a series of events during 2013 to 2014 which were well received by both industry and academia.

These events were delivered in partnership with the Scottish Textiles and Leather Association (STLA) and the Scottish Textiles Academic Group, as well as an introductory event related to the SCAP, delivered in partnership with WRAP. Matters discussed included the concept of the circular economy and the challenges and opportunities ahead for textile design and manufacturing at a national and global supply-chain level, as well as assessing the SCAP Commitment targets of 15% reduction in carbon, water and waste by 2020.

The STLA events also gave the industry the opportunity to share their experiences and ideas as to how to make the industry more efficient. This included a case study delivered by the Scottish Leather Group presenting their 11 years' experience of working towards sustainability.

Feedback from these events identified the need for training and skills development in resource efficiency, sustainability and global design and manufacturing models such as closed loop. Training bodies such as Skills Development Scotland and the Scottish Textiles Academic Group, and other sector funding bodies such as Scottish Enterprise, have welcomed the opportunity ZWS is offering the textiles sector, via its master classes, to introduce new skills and global textile knowledge on the circular economy.

The ZWS-commissioned research also provided evidence of related change in other nations. Specifically, ZWS asked the researchers to show how Scotland could learn from, and adapt, closed-loop models throughout the design, manufacture and supply chain and to identify if there were any early adopters or textiles businesses currently working towards such models. The researchers were tasked with reviewing and referencing previous research by partner organisations such as WRAP UK and the Ellen MacArthur Foundation, and ongoing academic research that would help ZWS identify key interventions it could make to embed circular economy and resource-efficient business models in the textiles sector in Scotland.

The brief for the researchers was to identify three key areas of research:

1) To provide a summary of the academic and industrial textile innovation landscape in Scotland, including developments in technical textiles as well as research into innovation in textile design in Scotland

2) To present case studies that reflect the landscape of Scotland's textile industry in relation to innovation in resource-efficient and waste reduction practices that are, or could be, developed as models of good practice in the circular economy

3) To present recommendations for future areas of support in terms of resource efficiency and closed-loop textiles in Scotland

The research was required to conclude with recommendations for future engagement and demonstrate how ZWS can help shape the future landscape of textile innovation in Scotland in relation to the circular economy and closed-loop textile manufacturing. The research needed to align with, and reference evidence of, the global trends and examples of current closed-loop practice and circular economy textile models in other countries, including the UK, that could be applicable to Scotland.

The report's recommendations have informed ZWS's work programme 2014 to 2016 for the textiles sector on how to shape the future landscape of textile innovation in Scotland, and it offered relevant examples of the circular economy from Scandinavia and advised developing links with work being undertaken there. It identified nine companies as potential case studies. This paper presents four of those nine case studies and outlines developments with Scandinavia.

With the exception of one company, Kraft Architecture, whose objective is to recycle waste textiles into cavity wall insulation, no true examples of circular economy business models in Scotland were identified by the researchers. They then proposed that the project scope be broadened to

Europe, arguing that the business environment is similar in landscape to Scotland and that any activities discovered might be transferable to the Scottish textile industry. They identified too ongoing projects that are exploring a number of initiatives, including producing an alternative to denim, an initiative first supported by WRAP UK and now being researched at Heriot-Watt University, Edinburgh.

2.1 The four case studies

Dawn Ellams - PhD Candidate, Heriot-Watt University

This research was sponsored by Lenzing AG to make 'denim' from TENCEL (Lenzing AG, Lenzing, Austria) using digital printing to impart the stone-washed look[j]. Given the high environmental impact of stonewashing denim, this could be a promising circular economy project if it is possible to commercialise it within Scotland. At present, the denim supply chains are located, typically, in countries such as Turkey, China or Pakistan. This work and research was further funded by WRAP through the Challenge Fund.

Jeans were created that have a cotton-like quality but only use one fifth of water, energy and chemicals needed to manufacture conventional jeans.

Manufacturing one pair of cotton denim jeans uses, on average, 42 litres of water and is energy intensive. Conventional denim production methods can also require up to 15 dyeing vats and an array of harmful chemicals [k].

The research identified several areas within the manufacturing process which offer opportunities for saving water and reducing carbon emissions.

Kraft Architecture - entrepreneur/inventor

Both virgin and waste wool and textiles can be converted into building insulation products, and there have been a number of attempts to set up businesses in Scotland supplying this as a product. Notable examples include Thermobond from Kraft Architecture (280 High Street, Glasgow

G4 0QT, UK) and Earthwool® from Knauf Insulation (PO Box 10, Stafford Road, St Helens, Merseyside, WA10, UK). Generally sustainable insulation products fail to compete on price with mineral wool, which is also principally derived from waste materials, and fail to compete on performance with engineered insulation materials. The resulting market is very small and tends to be limited to applications like historic property restoration where technical and commercial factors are less important than heritage.

Kraft Architecture is a sole practitioner company, although this is augmented by part-time help as the need arises. The business focuses on low-carbon and low-energy housing and sustainable product development. The company started in 2009 and was awarded a SMART: SCOTLAND Feasibility Award in 2010 to develop the product from the concept.

One of the closed-loop products developed by Kraft Architecture is a natural fibre insulation that retails under the trade name 'Thermobond'. The insulation manufacturing process can work with a wide variety of waste streams including cellulose, cotton, wool, wood and polyester and so it can create insulation materials tailored for different applications. A research project undertaken by Glasgow Caledonian University measured the thermal conductivity of a range of waste products to verify their potential for use as insulation[i]. All the selected materials compare favourably with glass fibre on this performance metric. Keeping the proportion of wool at 60% or greater ensures that product flammability and resistance to smouldering complies with the British Standard BS5803-4:1985.

The manufacture of Thermobond requires 90% less energy compared to typical mineral wool insulation. In use, its durability and moisture control properties allow for a higher level of building fabric performance over the life of the building. It is also soft to touch and non-irritant, making handling and installation simple. Thermobond is treated with non-toxic, naturally occurring boron salts to heighten its resistance to vermin and insects over the life of the product.

Samples of Thermobond have been produced in sufficient quantities for use in demonstration installations. Kraft Architecture has secured investment of about £500,000 from the recently created Construction Scotland Innovation Centre[m] and further funding is being considered by ZWS. This will include suitable machinery to make insulation materials from recycled materials in its equipment set. Whilst this will not provide a volume manufacturing facility, it will provide a route to further product development, as well as sufficient material for the occasional demonstration installation.

Turnberry Rug Works-Johnstons of Elgin - a closed-loop approach

Turnberry Rug Works and Johnsons of Elgin are working collaboratively to achieve the up-cycling of cashmere.

Established in 1991, Turnberry Rug Works, based in Ayrshire, is a small-to-medium enterprise, well known for its rug-making expertise. Its products are aimed at the higher end of the market, being both handmade and often bespoke. Although most of its output uses merino wool, it can incorporate stones, leather, seashells, steel and glass, as well as silk, linen, worsted wools and felted yarns into the designs.

Johnstons of Elgin is a long-established knitwear manufacturer (since 1797) that specialises in cashmere products. The company is vertically integrated, with the incoming feedstock being cashmere sourced exclusively from Inner Mongolia. The company has both a knitting plant in Hawick and a weaving plant in Elgin.

Weaving and, to a lesser extent, knitting processes, produce waste that cannot easily be recycled. A common type of waste is selvedge, which is the semi-finished edge that is produced to stop the fabric from unravelling or fraying. The production of scarves produces long lengths of this material. Following a chance meeting, Johnsons of Elgin supplied samples of its scarf selvage to Turnberry Rug Works which, after some development work on the tufting machines, found it was able to use the selvage to manufacture rugs. Producing a rug using virgin cashmere would be a very expensive proposition; however, by using selvage, a luxurious product can be produced using material that would otherwise have been sold as waste. The rugs are currently in production and sold under the Ecochic brand.

Keela International

Keela International Ltd specialises in the manufacture of advanced clothing for outdoor wear. Customers include consumers, mountain rescue teams, police and military forces. This is a large but highly competitive market sector estimated to be worth about €5 billion in Europe alone. Because both the USA and European markets are saturated, effectively, innovation is important to help sustain market share. Manufacturer claims that can be used to promote sales include 'free of toxins, environmentally friendly materials and sourcing, temperature control, reduced friction, water and wind resistant, moisture management'. The differentiating characteristics of Keela include expertise in fabric technology and garment construction.

Keela was awarded the Queen's Award for Enterprise: Innovation for its System Dual Protection fabric. This consists of two fabrics with very different characteristics, separated by an air gap. Originally developed for a military customer, it is now part of Keela's consumer product range. Independent research, conducted by Leeds University and Niederrhein University of Applied Sciences, substantiated the claims made for the material in terms of its ability to keep the wearer dry from external water and comfortable by regulating internal moisture.

Most modern outdoor clothing is made from materials that are difficult to dispose of, except in landfill; therefore, the most sustainable option is to maximise the life of the garment. This requires the customer to follow specific instructions on washing and reproofing and adds to the unfavourable environmental impact through the use phase of the textile.

As part of the company's drive towards sustainability, they have committed to SCAP and are the only apparel manufacturer member from Scotland.

2.2 International relations - Scandinavia

The researchers suggested to ZWS the need to explore case studies in Europe that might provide inspiration and examples for Scottish textile businesses to follow. Scandinavia was the prime suggestion, and a number of examples of the circular economy in textiles were identified.

The Research Council of Norway (NICE) funded a 3-year project (2010 to 2013) valued at €500,000, with a view to identifying and promoting the benefits of textiles made of Norwegian wool to local consumers. The budget was used to support a variety of promotional activities, including publishing a book, *Ren ull* (Pure Wool, Aschehoug). It is currently only printed in Norwegian, with a foreword in English by Prince Charles. Its approach is similar to the Campaign for Wool.

When the project ended, NICE wanted to continue its research and use the heritage and technology it had discovered to develop further consumer demand for Norwegian wool and create a label of provenance, similar to Harris Tweed, called the Nordic Swan Ecolabel. As part of this research, they chose to visit Shetland where fleece is gathered from over 700 crofters to produce Shetland wool.

Norway also developed a product which was given an international standard-approved Cradle to Cradle[n]. The Dutch airline, KLM, wanted to create Cradle to Cradle carpeting working with Desso. Norwegian wool was tested for chemicals, etc. and was approved. The research company brokered an introduction to NICE and ZWS and the potential to develop research collaborations around wool and revise the heritage of Scottish and Norwegian knitwear has been identified. In this way, each country can work towards a model of*provenance, traceability, durability* and *quality*, as per the four key values mentioned earlier that the ZWS-commissioned report recommends.

Another example is the 'Aran jumper', a well-crafted, distinctively patterned garment, synonymous with Scotland and Ireland, that is constructed so soundly, it can be disassembled, unravelled and reknitted, thus demonstrating an excellent closed-loop process. Families often deconstructed two small garments, knitting them into one larger garment, as their needs dictated.

The challenge today is to see how that heritage and evidence of closed loop can be harnessed to the circular economy to achieve a sustainable future for the textiles sector.

There is scope too to draw on the cultural craft heritage of Scotland; for example, there is a demand for hand-crafted products that draw on that heritage and once declining Scottish skills such as hand knitting, which have been invigorated by knitwear design companies such as Di Gilpin, who currently employs 90 freelance hand knitters throughout Scotland. Similarly, Eribe knitwear has 200 registered hand knitters who can each generate income of between £100 to £400 per month producing contemporary Scottish knitwear for export, mainly to China and Japan. Through the support from Scottish Enterprise, companies such as Eribe have doubled their turnover in 5 years, mainly through exporting.[o] These companies are challenged to source sustainable Scottish raw materials such as wool from Scotland (other than from Shetland) and a growing number of small sheep crofts. The textile sector has to be aware, however, that a revival in craft heritage matters has to consider existing industry frameworks. For example, since 1952, a sheep farmer with more than four sheep is required, legally, to sell wool only through the British Wool Marketing Board (BWMB). The penalty, if enforced, is still a 6-month jail term. In 1995, this legislation was relaxed to allow farmers to sell wool only for direct export. This was due to the low price, lack of demand and essentially oversupply, with BWMB not being able to clear the stocks.

3. DISCUSSION AND EVALUATION

The ZWS-commissioned research concluded that the Scottish textile industry is innovative in key sectors, particularly technical textiles and computer-aided design. However, in both cases, the overarching objective is either to increase business competitiveness or improve resource efficiency, rather than closed-loop sustainability. The report concluded that there is little innovation in business models or in new methods of making the sector more materially circular in ways that are commonly understood. It also concluded that Scotland has a strong technical textiles sector, and there may be specialist opportunities here for closed-loop recycling solutions.

It recommended that the Scottish textile industry should focus on strengthening what it already does well and bolster its sustainability activities by linking to *provenance, traceability, durability* and *quality*. Inspiration should come from the luxury goods sector and also, in a more modest way, from Scandinavian brands such as Filippa K, where longevity is an explicit brand strategy.

ZWS has integrated the report recommendations into its work plan to March 2016. It will focus on five key areas:

1) Industry and academia - skills training

2) New business model research

3) Fibre-processing facility

4) Funding to experiment with closed-loop technologies and systems

5) Textiles and apparel

3.1 Industry and academia - skills training

A series of master classes will be delivered from April to December 2015. The topics and content for the master classes are based on an analysis of the key characteristics of Scottish textiles design and manufacturing that will enable resource efficiency and skills for a circular economy such as design for disassembly and fibre processing and resource efficiency topics such as zero-waste pattern design and new technologies in dyeing and printing. One master class, Dyeing and Chemical Processing, will introduce global brand research such as dry dyeing by Stella McCartney for Adidas (Amsterdam, the Netherlands) and focus on resource efficiency such as a reduction of effluent from chemicals.

The majority of the 16 places for each course will be allocated to industry, with a quarter allocated to academia and education leaders who are developing curricula. Attendance is by invitation and application and candidates need to evidence how they will cascade the learning within their business or within departments and to colleagues.

Each facilitator will be travelling to Scotland from other parts of the UK or Europe (where possible, webinars will be used). To maximise the purpose of each visit, a student lecture series will also be part of the package accommodating up to 200 students at each lecture. Edinburgh University, Heriot-Watt University, Glasgow School of Art and Duncan of Jordanstone College of Art and Design in Dundee have provisionally agreed to host the these lectures.

Skills Development Scotland has confirmed that it is complementary to the Scottish Textiles Skills Action Plan[p] that has recently been funded by the Scottish Funding Council.

ZWS will source and invite experts from outside Scotland, and evaluation consultants will be appointed to assess the impact of the participants' learning experience and how they have cascaded their learning to other colleagues or implemented new skills in their business processes.

3.2 New business model research

ZWS has identified two textiles companies who would like to explore new business models, details of which will be published in 2015[q].

An example of a business model that was identified in the research and that has characteristics that could apply to Scotland is Waste to Wow! Established in 2013 and based in Italy, it is a start-up enterprise that has achieved the accolade of being winner of the 2014 European Social Innovation Competition, out of 1,254 entrants[r]. Their take on recycling is slightly different in that their waste stream is textiles from high-end fashion manufacturers that have failed quality control checks. Normally, these garments would be destroyed to make sure there is no leakage onto the grey market. From Waste to Wow! assures confidential and controlled disposal by using the material to make new, limited edition textiles that they sell under the brand name Quid (Progetto Quid,

Verona, Italy). Thus, their feedstock tends to be of much higher quality than most similar recycling operations based on this model.

Scottish textile manufacturers who supply luxury global brand textiles must destroy their seconds and textile waste to avoid grey market leakage. An interior textiles company and some apparel textile companies have asked ZWS for advice as to how to dispose of luxury brand seconds currently going to landfill or being incinerated.

3.3 Fibre-processing facility

ZWS is interested in exploring a fibre-processing facility in Scotland that would help tackle the large fraction of lower grade bulk textiles which currently go to landfill, including carpets, mattresses and post-industrial waste from the textile industry. It also offers the opportunity to develop investment and jobs in fibre reprocessing in Scotland, adding value to materials which are currently worth little. As an example, if 60% of mattresses can be recycled (7,200 tonnes equates to 335,000 units), direct jobs created are estimated at 80 based on the ZWS business case of 8 jobs per 34,200 units.

A fibre-processing facility would enable Scotland to process post-industrial and post-consumer products from throughout the UK and gain some of the market share that other areas of the UK benefit from. Cotton and polyester are, respectively, around half and one quarter of the UK clothing fibre mix and hence are the priorities for fibre processing. Mixed fibres such as wool and nylon are characteristic of carpet construction, and mixed fibre extrusion is still a developing area. In the case of carpets, Anglo Carpets in England employs 35 people to process 1,200 tonnes. Carpet Cycle employs 50 people for 5,500 tonnes. Mid-point is one person per 75 tonnes processed.

The WRAP report, *UK textile product flow and market development opportunities*, has numerous recommendations regarding how the UK textiles market could be developed. ZWS' priority for this work, based on evidence in that WRAP report, is to investigate the potential for Scotland to reprocess cotton and polyester back to fibre or feedstock, at cost parity to virgin material. It is recognised that scale could limit the implementation of the reprocessing infrastructure and this may require a change of direction for this area of work.

The research areas for the fibre-processing facility are specific to the following:

- Identifying candidate technologies for fibre processing and identifying their needs for financial support to enable proof of design, trialling and scale-up
- A route map of the activities which might be needed to deliver a commercially viable service

Following the identification of suitable technologies and their applicability to Scotland, a financial and economic assessment will need to be conducted.

3.4 Funding to experiment with closed-loop technologies and systems

Funding will be offered by ZWS to three projects through an open call for industry and academic partnership projects. Project proposals should explore the feasibility of a new textile product or process that demonstrates closed-loop design or manufacturing. Awards will favour technical and performance textile companies.

Projects should be completed in January to December 2015. They will be monitored and evaluated externally to capture shared learning and secondary outcomes that may not be evident from the trialling of a product or process. Funding will be allocated for external monitoring and evaluation. Those projects which prove feasible maybe further supported by ZWS and taken to prototype/trial stage.

There will be *three* themes:

1) Closed-loop manufacturing - companies that already manufacture textiles and have post-industrial waste going to landfill that they think has the potential to be reprocessed

2) Piloting technologies/products for disassembly - companies that would like to trial technologies such as Wear2 microwave seam technology

3) Exploring new fibre opportunities - all proposals will be considered from recycled synthetic fibre to the introduction of large-scale natural fibre processing

3.5 Textiles and apparel

Scotland has a strong fashion and textile design community, with 600 textile and fashion graduates every year and graduates such as Lauren Smith from Edinburgh College of Art winning the prestigious Gold Award at London Graduate Fashion Week 2013. ZWS would like to encourage designers working in Scotland to consider supply chain sustainability and research textile design and garment construction that relates to circular economic and resource-efficient practices in design such as design for disassembly and sourcing and designing with recycled yarns such as recycled P.E.T., as well as alternative fibre and plant dying processes.

ZWS will provide funding for designers to develop prototypes working towards a truly sustainable circular economy product and/or using alternative fibres to imported yarn with a heavy carbon footprint. Mentors will be provided for each winner. The challenge will be finding suitably knowledgeable mentors in Scotland in this field.

4. CONCLUSIONS

Given the increasing demand for Scottish textile products, ZWS' work programme 2014 to 2016, and recent discussions with NICE, Norway, the Scottish Government's policy for the textiles industry should consider the following:

- Does Scotland need its own sustainability standard? Will this help or hinder business and how important is it to the provenance of Scotland's textile industry?
- How viable are pilots with the Scottish textiles industry to introduce new technology trials that will develop closed-loop processes?
- Can we source international investment in Scotland that not only provides job security but also invests in new technology and skills for a circular economy?

ZWS' support for the textiles sector in Scotland is already established. The key challenge now is to take forward the five aforementioned action points and evaluate and evidence the impact of embedding a circular economy in those areas. This paper has outlined our approach, why we have chosen it, and how we plan to implement it. The specific context is the potential for business growth in the textile sector in Scotland, within the broader context of ZWS' support for the Scottish Government's commitment to its zero-waste targets for 2025. Measuring the impact of a circular economy approach is a task for the long term and we know from industry examples, such as TENCEL® by Lenzing[s], that it can take up to 30 years to develop a product that revolutionises industry.

The production of TENCEL® is revolutionary. The production process is based on a solvent spinning process and represents the greatest accomplishment in cellulosic fibre technology. The unique closed loop production process makes TENCEL® the fibre of the future: eco-friendly and economical[t].

Options for development and investment in Scotland, therefore, not only rely on ZWS and Scottish Enterprise and other development agencies but also require collaborations and investment from the rest of the UK, including further and higher education, and global partners. ZWS' work programme from 2014 to 2016 will produce evidence and provide incentives to the textiles sector that show our commitment to underpinning a circular economy.

ZWS is intent on supporting a sustainable industry that will generate long-term growth and be supportive of the Scottish Government's zero-waste targets. That can only come about by facing up the challenges and opportunities that a circular economy model, as per our work programme for 2014 to 2016, is intent on discovering. ZWS will maintain alliances with all the key stakeholders mentioned in this paper, as well as forge stronger connections with the industry and academia. In our view, the theory and practice engendered by that connection can establish a body of knowledge on the circular economy that will determine the Scottish textile industry's short-, medium- and long-term goals and secure its future sustainability.

ENDNOTES

[a]Scottish Textiles Strategy 2011 to 2015

[b] http://www.ellenmacarthurfoundation.org/

[c]The Scottish textiles sector has been given £750,000 over 3 years to develop innovation and skills for industry. Heriot-Watt University, Edinburgh will administer the programme. Glasgow Clyde College will deliver an academy-style skills programme funded by a consortium of funders including Skills Development Scotland, and ZWS will be invited to a sub-group meeting when it is set up.

[d] http://www.gov.scot/Publications/2013/10/6262. Exploring the feasibility of circular economy business practices contributes to 'Safeguarding Scotland's Resources', Sections 6.1 Resource Efficient Scotland - Action 7 - Business engagement and an early adopter network and Action 12 - Remanufacturing and new business models, 6.2 Stimulating innovation and business opportunities and the provision of technical master classes for the textile design and manufacturing sector

supports and 6.3 Sustainab`le Design Action, Objective - to stimulate innovation in the design and manufacture of products and packaging - to engender a shift towards design for longevity, reassembly or remanufacturing and useable by-products, designing out wasted energy, materials and pollution. In addition, by inviting teaching staff in education and academia to participate in the master classes with industry, it also supports Action 18 - Public engagement and education. ZWS works closely with Education Scotland, Young Scot and the Ellen McArthur Foundation on these activities

[e] http://www.ellenmacarthurfoundation.org/about/media - Richard Lochhead, Environment Secretary, statement on joining CE100 - 'Scotland's economy will benefit from moving to a more circular model of production and consumption. Our Zero Waste Plan is already delivering important actions to make better use of resources, and we can accelerate progress if we join together with others on a global level'

[f] http://www.knittingindustry.com/chanel-to-create-100-new-jobs-at-barrie-knitwear-in-scotland

[g] http://news.scotland.gov.uk/News/Textiles-target-success-564.aspx

[h] Scottish Key Facts, compiled by the Economic Research Team, Scottish Enterprise, May 2014

[i] http://www.itv.com/news/border/story/2013-08-07/mongolian-delegation-in-hawick/

[j] http://www.hw.ac.uk/news-events/news/wooden-denim-to-cut-carbon-emissions-11090.htm - Dawn Ellams, Feb 2013

[k] http://www.hw.ac.uk/news-events/news/wooden-denim-to-cut-carbon-emissions-11090.htm

[l] http://www.cicstart.org/userfiles/file/AC-1-REPORT.PDF

[m] http://www.gcu.ac.uk/newsevents/news/article.php?id=73608

[n] http://www.c2ccertified.org/product_certification

[o]Scottish Enterprise case study

[p]The Scottish Textiles Skills Strategy & Action Plan has been developed to support the objectives of the industry strategy *A Strategy for the Textiles Industry in Scotland 2011-15 (* http://www.scottish-enterprise.com/knowledge-hub/articles/publication/textiles-strategy-2011-2015 *)* and sets out the key skills priorities that the textiles sector must address in order to fulfil its growth potential

[q]New business models are covered by a wider stream of the Circular Economy Team and therefore, were not part of the commissioned textiles research, although there are relevant documents within the report

[r] http://socialinnovationcompetition.eu/401/

[s]The Lenzing Group is a large publicly owned manufacturer of cellulosic fibres. It employs over 6,500 staff, making nearly 900,000 tons of fibres per annum, with sales of around €2 million per annum. Its three principal products, Viscose, Modal and Tencel are well known to the sustainable textile community. The feedstock is wood, sourced from sustainable forests. Fibre production by Lenzing achieves a wood utilisation rate of over 50%. Co-products include furfural, xylose and acetic acid with the remaining parts used principally for energy production. In 2013, Lenzing launched a new fibre, EUROCEL, which is a 50/50 blend of Viscose and Tencel. The intended application is high-end non-wovens, mainly for hygienic applications as well as in the household and industrial sectors

[t] http://www.lenzing.com/en/fibers/tencel.html

6

ENVIRONMENTAL SUSTAINABILITY THROUGH TEXTILE RECYCLING

Chavan RB*

Institute of Sustainable and Appropriate Technology, Aurangabad, India

ABSTRACT

The paper examines the process of achieving environmental sustainability through recycling of textile wastes. The recycling of textile waste can serve as a means of providing solutions to many economic, environmental and social issues. Though textile recycling has old history; in recent years it has assumed prime importance due to Fast Fashion culture in western world which has resulted in over consumption of textiles and corresponding waste generation. Today waste recycling has become a multibillion industry. Innovations are being made in terms of development of sorting machines, design inputs, innovative high value products to make recycling a profitable proposition. The paper also highlights in brief the efforts being made by various agencies and stake holders to achieve the aim of present 15% textile waste diversion rate to 75% diversion.

KEYWORDS

Reuse; Recycling; Environmental sustainability; Waste diversion; Global warming; Climate change; Resource consumption

1. INTRODUCTION

For most of human history the world was sparsely populated and resources from forests to fisheries seemed unlimited. In the 20th century there were 1.65 billion people in the world, a population not much greater than that of China or India today. Society used resources and generated waste with little restraint and few concerns. This helped drive growth and improve the livelihoods of many millions of people.

The global population has now reached 7.2 billion and is estimated to be 9.6 billion by 2050 [1]. Resources are becoming increasingly scarce and climate change is a reality. This will have severe effects on weather patterns, water availability and agriculture [2]. There is loss of half the world's forests and degradation of an estimated 60% of the world's ecosystems [3]. When it comes to the environment, society is currently using resources at a rate that requires 1.5 planets [4]. These burning environmental problems arise from ever increasing volumes of worldwide production and consumption and the associated material flow waste [5].

Efforts are being made to create awareness for the catastrophic situation and educate people to reduce consumption and recycle products. Though recycling rates have increased around the world, but the majority of valuable processed materials are still thrown away rather than reused or recycled, leaving room for significant improvements and opportunities. Rising energy and raw material costs are putting pressure on businesses across the world. In the absence of bold decisions, price increases for energy, water, raw materials and manufacturing activities including textiles cannot be avoided. It is essential to understand that the 20th century approaches to meet 21st century demands are not affordable for sustainable development. Thus it is essential to incorporate the philosophy of sustainable development as an integral part of economic and environmental policy not only in developed countries but also in developing countries.

In view of the above futuristic view for sustainable development and business opportunities it is but natural to reduce the consumption of resources through recycling of products which are considered to be waste Textile recycling has old history. The concepts of recycling, together with reduce & reuse have been ingrained in the social milieu of India. Traditional societies had in-built models of sustainability, through reuse and recycling. Production and consumption remained in harmony with each other and remained largely need driven [6].

In China, more than 2000 years ago, used clothing was shredded and hand-carded for blending with virgin fibers to make yarns. During the Napoleonic war, a shortage of wool for military uniforms advanced the need for recycling postconsumer fibers. In the United States, textile recycling is one of the oldest, yet most misunderstood, recycling industries [7].

Prior to World War II, the United States was a major importer of rags, cuttings, and other textile waste materials because the United States was a leader in mechanization for converting the used textiles into manufactured raw materials. The situation has been reversed since World War II. Now the United States exports tons of textile waste materials primarily because of the overconsumption of textile goods that eventually results in textile waste [8].

1.1 Fashion as a Source of Over Consumption

Globalization has made it possible to produce clothing at increasingly lower prices, prices so low that many consumers consider this clothing to be disposable [9]. Some call it "fast fashion", the clothing equivalent of fast food. Development of technology has enabled the industry to produce a plethora of products resulting in the depletion of natural resources and indiscriminate disposal habits [10].

Thus, fashionable goods contribute to consumption at a higher level than need. As consumers continue to buy, waste will continue to be created, further compounding the problem of what to do with discarded waste apparel, and home textile products. This presents a double-edged sword, in that while at the same time it stimulates the economy, it also gives rise to the increased problem of apparel and textile disposal [11]. Nonetheless, the recycling industry must cope with everything that the fashion industry has generated.

1.2 Types of textile waste

Like all wastes, textile waste originates from a number of streams including the fiber, textile and clothing manufacturing industry, consumers, the commercial and service industries [12].

There are three types of textile wastes namely, pre-consumer, postconsumer and industrial.

Pre-consumer textile waste: Pre-consumer textile waste is the manufacturing waste that is generated by processing fibers, (natural or synthetic) and the production of finished yarns and textiles, technical textiles, nonwovens, garments and footwear, including off-cuts, selvages, shearings, rejected materials and/or B-grade garments, offcuts of saleable size etc. Pre-consumer textile waste is usually "clean waste". Firms either arrange their own waste disposal services or use council managed services and pay landfill fees according to how much is dumped.

Post-consumer textile waste: Post-consumer textile waste consists of any type of garments or household textiles (such as sheets or towels) that the consumer no longer needs and decides to discard, either because they are worn out, damaged, outgrown, or have gone out of fashion. This category has typically been of reasonable to good quality garment that can be recovered and subsequently recycled by another user as second-hand clothing, much of which is sold to third-world nations. Clothing that is unlikely to be worn again; may be shredded into fiber to be used in products similar in nature to those manufactured from pre-consumer textile waste.

Industrial textile waste: Industrial textile waste is generated from commercial and industrial textile applications including commercial waste from properties such as carpets and curtains, hospital refuse in addition to industrial applications such as filtration, conveyor belting, etc. Industrial textile waste is usually "dirty waste". Collection and chemical contamination issues render this category as the least likely to be recovered. A substantial proportion of these end-of-life goods are incinerated or dumped to landfills.

However, there is research currently being undertaken by a number of industries, including the carpet industry, to utilize this resource.

1.3 Recycling of textiles waste

The detail discussion on recycling of textile waste is out of the scope of present paper and is restricted to the point of environmental sustainability. The textile waste primarily can be recycled in the form of reuse and conversion into low value or value added products. Although the present emphasis is on recycling for value addition which is commonly known as up cycling.

The importance of recycling of textiles as a source of reduction in the use of virgin fibers and other resources such as water and energy can be understood from the following typical facts.

Environmental impact of cotton and cotton/polyester t-shirt

The resource consumption for cotton and cotton/polyester T-Shirt and CO2 saving are given in Tables 1 and 2 respectively [13].

Greenhouse gas savings from textile recycling

For every pound reused or recycled, textiles account for more greenhouse gas savings than paper, plastics and glass combined as shown in Figure 1.

Textile waste generation and diversion rate

In USA; by 2015 annual textile waste generation will increase to almost 14 million tons. Diversion is stagnant at 15% as illustrated in the following Figure 2 [13].

The aim is to achieve 75% diversion.

The enhancement of diversion of textile waste to convert into value added products could be achieved by adopting a proper strategy of

1. Increase convenience

2. Increase awareness

3. Measure the results

Convenience

Convenience of location and access is important when diverting clothes for recycling and reuse. Some of the efforts needed to increase convenience are

- Outdoor recycling bins
- In-store collection points
- Corporate collection
- Multi-family units
- Transfer stations
- Public facilities - parks, libraries, fire stations
- Schools Universities and prayer halls
- Curbside collection
- Events

Increase awareness

- Government, non-profit and private enterprise
- Partnerships can help
- Educate the public on textile recycling.

Measure results

Include textiles in diversion planning and reporting.

The effect of environmental impact of 75% textile waste diversion

The effect of environmental impact of 75% textile waste diversion in terms resource saving, economic, and social issues are listed below

Resource saving:

- 4.2 trillion gallons of water would be saved. That's enough to supply 27.8 million homes.
- 17 million tons of CO_2 would be saved. That's equivalent to taking 3.5 million cars off the roads.
- 7.5 million cubic yards of landfill space will be saved. That could fill the Empire State building 5.8 times!

Economic

Increasing textile diversion to 75% would create many employment opportunities.

Social

70% of the world's population wears second hand clothing.

International efforts to enhance diversion of textile waste

The vision of every organization and municipality is to recycle all of its waste into usable products, there by closing the life-cycle loop. Science teaches us that matter and energy are not destroyed, they merely change state. Waste contains significant amounts of valuable resources that were once used as a raw material. Those products have lived their useful lives and have become waste, but they still contain the same matter end energy that went into their making. Logic suggests that by recovering textiles, the demand for virgin resources is reduced. Material washing and energy consumption still occur in the recycling processes, but they are considerably less resource intensive and polluting than the processes involved in manufacturing textiles from virgin fiber. Various stages of the recycling process also provide significant employment around the world.

Today, recovering textile waste is a multi-billion dollar global industry that performs a vital social and environmental function and provides employment for millions of people all around the world. An internet search on "textile waste" will elicit thousands of products or listings, including headings such as hosiery cuttings and clips, polyester tow, cotton shoddy, used clothing wiping rags, denim/jean clippings, 100% cotton yarn waste, silk fiber waste, etc.

All textile waste streams are often unrealized sources of valuable raw materials that can be repurposed or regenerated into saleable and usable products by intelligent collection, sorting, re-engineering and reprocessing. In essence, the liability of "waste" is turned into an asset often based on intellectual property (IP) which has been specifically developed. Processing machinery is also likely to be engineered to produce a specific product and thus the investment in developing regenerating capabilities is often large-scale. Products made by regenerating textile waste include acoustic textiles used for sound proof blocks, insulation, roofing felt, bank stabilization, and as pollution control filters. The list can be left to the imagination.

Some of the international efforts to divert textile waste into valuable products are only briefly mentioned. The details could be obtained from respective website references. The list is by means complete. The limited listing is indicative of the seriousness in the business of textile recycling for environmental sustainability.

1.4 Commercial Organizations

Textile for Textiles (T4T)

The quality of the recycled output depends on the accuracy of the sorting process. The T4T project uses an automated system which can sort textile waste, based on composition and color, quickly and accurately. An automated industrial sorting line using the NIRspectroscopy has been developed and demonstrated. Having analyzed the economic benefits, T4T upgraded and adapted its installation based on sorting requirements in relation to the envisaged products from the recovered fibers. Sorting system can be installed at a post-consumer textiles and clothing sorting plant. The installation will be able to sort large quantities of textile materials based on fiber composition (cotton, wool, polyester, blends, etc.), color, structure, coating or finish composition [14].

Patagonia common threads recycling program

Patagonia's 'Common Threads Recycling Program' uses a polyester fiber-to-fiber recycling system to make new garments from old. Patagonia has partnered with Japanese textile firm Teijin to implement a polyester product recycling program which started in Aug 2005. Under the program, Patagonia will recover from customers used Capilene undergarment products. Teijin will recycle them as polyester materials at a "fiber-to-fiber" recycling facility in Teijin Fibers' Matsuyama plant. In 2005, Patagonia began using a new polyester filament yarn containing 30%-50% post-consumer feedstock made from discarded polyester bottles, polyester uniforms, tents, and garments.

On the environmental benefits Patagonia states: "Research shows that the environmental impact of using worn-out clothing to make new polyester fiber is significantly lower than making the same fiber from virgin materials derived from petroleum. By diverting worn-out garments from landfills, we reduce solid waste. Diverting them from incinerators results in an energy savings of 76% and a CO2 emissions reduction of 71%" [15].

Extended producer responsibility

In the field of waste management Extended Producer Responsibility (EPR) is a strategy designed to promote the integration of environmental costs associated with goods throughout their life cycles into the market price of the products [16].

The concept of first introduced in Sweden with the following definition. EPR is an environmental protection strategy to reach an environmental objective of a decreased total environmental impact of a product by making the manufacturer of the product responsible for the entire life cycle of the product and especially for the take-back, recycling and final disposal [17].

Take back program

Many brands and retailers have shown interest in this issue and have started to take ownership of their products beyond retail, developing take-back programs with incentives.

It would not be impossible to name all current initiatives but here are a few examples:

Klattermusen: Swedish firm has launched the program rECOver that allows its used products to be returned to retailers, for cash [18].

Mammut: Swiss firm is providing collection boxes in its stores in Switzerland and partnering with a collector who is sorting the products and reusing or recycling them as much as they can [19].

Marks and Spencer: Since 2008, Marks and Spencer and Oxfam have been developing a cloth exchange program. Since then, they have managed to collect over 2,500 tons of clothing which are put to reuse or recycled [20].

On line market places

RecycleMatch: RecycleMatch [21] is an online marketplace to address the $90 billion market for business recycling and zero waste with a closed bidding system designed specifically for corporations, municipalities, recyclers, and buyers of high quality commercial waste, in order to maximize the economic value of zero waste initiatives. There are hundreds of other exchange networks for industrial and municipal waste. Some of these are organized by non-profit organizations, government, or commerce chambers. They are free or inexpensive for companies to use them. Others are run by specialized companies and financed by a fee for announcement, or are consultancy providers who provide advice to waste producers on potential applications of their waste streams. Some are local, some regional or national, and some international. One can do an online search for 'waste exchanges' to find the one that meets your criteria. Some of the main ones include: Global Recycling Network [22], Euro recycle net [23], Worldwide Recycler's exchange [24].

Government organizations/Associations/Institutes

Sustainable Clothing Road Map: The UK government's Sustainable Clothing Roadmap launched in 2007, with the aim of improving the sustainability of clothing across its lifecycle from the crops that are grown to make the fabrics, to the design and manufacture of the garment, retail, use and end of life [25].

Sustainable Clothing Action Plan (SCAP 2020): Signatories and supporters of the SCAP 2020 commitment have pledged to play their part in reducing the carbon, waste and water footprints of clothing they supply or receive in the UK by 15% starting from the baseline year of 2012. 56 organizations have signed up to the commitment. Signatories and supporters represent more than 40% of UK retail sales, collectors, recyclers and charities [26].

BIR: The Bureau of International Recycling represents over 750 member companies from the private sector and 40 national associations in more than 70 countries. Together, these members form the largest international recycling federation. One of their expertise divisions is the industry of the textiles commodities [27].

SMART: Founded in 1932, the Secondary Materials and Recycled Textiles Association is composed of companies from the United States, Canada, Mexico, South and Central America, Europe, and PacificRim countries who are involved in every phase of the textile recycling industry: Reclaiming, converting, and recycling textiles and secondary materials [28].

Green Growth: New methods for recycling plastic and textile waste

Finding new methods to process waste and utilize it as a resource is among the Nordic Prime Ministers' green growth priorities. A new initiative, Resource Efficient Recycling of Plastic and Textile Waste, has been launched. The aim is to ensure that reuse will be given top priority. The objective is to develop Nordic waste collection and waste treatment methods, as well as new business models that contribute to green growth and can be exported to the rest of the world. In addition, the project will develop an Extended Producer Responsibility model, specifying producer's involvement in the effort to recycle their products [29].

2. CONCLUSION

The issues due to textile waste generation are related to over consumption of textiles in western world. The diversion of old textiles for reuse meets the requirements of clothing for 70% of the population in developing countries. Like other wastes, textile waste disposal through landfill and incineration is responsible for global warming. Though textile recycling has old history, today it has become multibillion industry producing innovative high value products. Many agencies and stake holders are making efforts to contribute towards the aim for saving in resources of raw materials, energy water etc. and reducing the impact on environment for sustainable development. In future textile recycling would be as important industrial activity as textile manufacturing.

REFERENCES

1. UN report (June 13, 2013) World population projected to reach 9.6 billion by 2050.

2. Inter-governmental Panel- IPCC (2000) Climate Change Special Report on Emission Scenarios.

3. Prosperity without growth (2011)The transition to a sustainable economy, Sustainable Development Commission.

4. WWF (2010) Living Planet Report.

5. UNEP (2007) Global Environmental Outlook 4.

6. Varsha G (2012)AjitKhareValue Creation in post-consumer apparel waste: a study of urban-rural dynamics in India. NIFT, India.

7. Nousiainen P, Talvenmaa-Kuusela P (1994) Solid textile waste recycling Paperpresented at the Globalization–Technological, Economic, and Environmental Imperatives, 75th World Conference of Textile Institute, Atlanta, USA.

8. Jana MH (2006) Digging for Diamonds: A Conceptual Framework for understandingReclaimed Textile Products. Clothing Textiles Res J 24: 262-275.

9. Claudio L (2007) Waste couture: environmental impact of the clothing industry. Environ Health Perspect 115: A448 - A454.

10. http://en.wikipedia.org/wiki/Fast-fashion

11. Katkar PM, Bairgadar SM (2010) Textile waste recycling Textile Review.

12. Caulfield K(2009) Sources of textile waste in Australia. Apical International, Australia.

13. WallanderM (2014) Increasing the diversion of textiles.USagain.

14. www.textiles4textiles.eu/

15. http://www.patagonia.com/us/common-threads

16. OECD (2001) Extended producer responsibility: A guidance manual for governments Paris, France.

17. Thomas Lindhqvist (1993)The concepts of extended producer responsibility and product. USA.

18. www.klattermusen.se

19. http://www.mammut.ch/cr_planet_ico.html

20. http://www.oxfam.org.uk/shop/content/

21. http://www.recyclematch.com/landing/homelandingpage

22. http://www.grn.com/a/1400.html

23. http://euro.recycle.net/

24. http://www.recycle.net/

25. www.defra.gov.uk/publications/files/pb13461-clothing-actionplan-110518.pdf

26. www.wrap.org.uk

27. www.bir.org/industry/textiles

28. www.smartasn.org

29. http://nordicway.org/2013/10/new-methods-for-recycling-plastic-and-textile-waste/

7

MATERIALS AND MANUFACTURING ENVIRONMENTAL SUSTAINABILITY EVALUATION OF APPAREL PRODUCT: KNITTED T-SHIRT CASE STUDY

Md Mashiur Rahman Khan and Md Mazedul Islam*

Department of Apparel Engineering, Bangladesh University of Textiles, Tejgaon Dhaka-1208, Bangladesh

ABSTRACT

In this paper, evaluation of base materials and manufacturing environmental sustainability of selected branded T-shirt products made by Bangladesh is presented. The study is based on performing various eco-friendliness tests for the product base materials and evaluation though a standard tool named Higg Index. For this study, selected apparel branded T-shirt products from renowned brand, namely B, C, D, E, and F, and a researcher-developed BUTex-Innovation brand A are taken into consideration. Better environmental sustainability results, i.e., higher score, is obtained by A, B, C, and D branded T-shirts, but local branded product of E and F resulted in poor environmental sustainability with lower scores in terms of product eco-friendliness tests and High Index assessment tool. Besides, many weaknesses and opportunities for improvement of environmental sustainability in materials and manufacturing stages are identified and suggested to lead the textiles and clothing sectors towards sustainable growth.

KEYWORDS

Clothing products, Base materials, Knitted T-shirt, Higg index, Environmental sustainability

1. BACKGROUND

Sustainability is the major concern in the age of the modern world. For the textile and apparel sector, it has been a burning issue for many related concerned bodies. Challenges faced by the apparel manufacturers and retailers, brands with many incidents such as the Rana Plaza building collapse, the Tazreen garments fire tragedy in Bangladesh, and the crisis in Cambodia, led the clothing industry giants, nonprofit organizations, fashion designers, manufacturers, and consumers to be well aware towards sustainable textiles and clothing business practices. Over the past few years, increasing awareness of the environmental and social concerns surrounding the fashion industries and consumers has led to a rise in the implementation of sustainability initiatives. There has been a growing concern over apparel brands in improving their environmental impact and the social responsibility throughout their supply chains (Allwood et al. 2006). Most of the textiles and clothing product used by the consumer has a significant impact on the environment. For example, according to a study, one 0.5-lb cotton T-shirt requires 700 gallons of water during processing, uses 0.2 lb of fertilizers and 0.01-lb pesticides, and emits 6 lb of CO_2, 1.2 lb of fossil fuels, and 0.11 lb of other gases. One 0.5-lb pound T-shirt does 18.3 pounds of CO_2 emission on average from washing and dyeing 50 times in US (Fletcher and Grose 2012; USagain, 2015). Many consumers today do not know the extent to which these products impact the environment—low or high. They must account for the environmental impacts of the materials she uses, the resulting waste of the forms she chooses, how products are produced and packaged, where they will be made and then sold, causing the energy use for transportation, how consumers may use and dispose of the product (Adams and Frost 2008). A product is thought to be eco-sustainable if it is manufactured, used, and disposed of in a manner that would not harm the environment when compared to a product that was manufactured and used without any environmental concern. However, consumers have become more conscious of these impacts and are spelling out their preferences for eco-friendliness, thereby forcing the manufacturer to adopt clean technologies all along the supply chain to produce environmentally friendly products (Challa 2014). Environmental sustainability refers to the ability of something to continue without upsetting the earth's ecological balance. Sustainable apparel products can be defined as a part of the growing design philosophy and trend of sustainability, the goal of which is to create a system which can be supported indefinitely in terms of environmentalism and social responsibility (Adams and Frost 2008). Environmental sustainability in business refers to longevity but in terms of which natural resources the production process might draw upon, how resources are used and replenished, the overall impact of the final production on the environment, and where the product ends up following its disposal. The textile and apparel products impact the environment at every point along the product's lifecycle (Fletcher 2008). However, rapid growth in the textiles and clothing industry sector is playing a vital role in the economy of Bangladesh. Despite the growing necessity of environmental practices in the apparel industry, little research has explored methods and standard tools to guide apparel designers, manufacturers, merchandisers, and consumers in adapting environmental issues in their works. Many designers do not recognize how their designs impact the environment. The role of the designer must expand dramatically in light of current environmental concerns with the contamination and destruction of the ecosystem (DEFRA 2010). Today's designer, manufacturer, merchandiser, and consumer must be well informed on a vast array of topics, from methods of production to governmental regulations to life cycle analysis. They must account for the environmental impacts of the materials' uses, the resulting waste of the forms they choose, how products are produced and packaged, where they will be made and then sold, the causing energy use for transportation, and how consumers may use and dispose of the product (Goworek 2011). All products that are manufactured cause environmental degradation, either during manufacture, use, or disposal. This can be evaluated by looking at the different phases of the product's life cycle and taking action at the phases where it will be most effective to reduce the environmental impact. Among many techniques to study the eco-impact of products, the life cycle assessment is most widely used one. LCA examines products from initial stage to the final stage, covers the entire life cycle, and also examines eco-impact during the whole lifetime (Muthu et al. 2012). However, the life cycle of a product is long and complicated, covering many areas with many people involved in each phase. A remedial measure or policy may not be possible to address this aspect, but a variety of voluntary and mandatory tools will help to achieve this objective. These include eco-friendly tests for related clothing products, economic instruments, bans on certain substances, environmental labeling, voluntary agreements, and product design guidelines (IPP 2014). Hence, there is a growing need to evaluate the environmental sustainability of apparel products by standard methods, tools, and techniques to minimize the environmental impact. The Higg Index by the Sustainable Apparel Coalition is such a standard tool by which the environmental sustainability of apparel products could be evaluated. This study attempts to identify and analyze a number of these issues related to apparel product sustainability focusing on materials and manufacturing environmental sustainability by taking several local and foreign branded knitted T-shirts into consideration as a case study.

2. LITERATURE REVIEW

The clothing industry is a popular industry among consumers, but it has great effects on many environmental, social, and governance concerns (Challa 2014). The textile industry prepares the base materials and the fashion, and the apparel industry converts these materials to suit the desires and needs of consumers; both industries are responsible for high utilization of energy, water, chemicals, and resources from cotton to petroleum (Niinimäki and Hassi 2011). The poor on-site conditions of the textile factories and working environment have caused many problems for the workers and operators, forming the basis for social reforms. In addition, the precarious supply chain upon which many manufacturers rely to develop apparel products can cause many problems for merchandisers and retailers for sustainable apparel product manufacturing.

2.1 Sustainable apparel product

Sustainability is defined as the design of human and industrial systems to ensure that humankind's use of natural resources and cycles do not lead to diminished quality of life due either to losses in future economic opportunities or to adverse impacts on social conditions, human health, and the environment (Lewis and Chem 2006). Environmental science defines sustainability as the quality of not being harmful to the environment or depleting natural resources and thereby supporting long-term ecological balance. The sustainable textile products can be defined as the textile products produced using raw materials, energy, various resources, and other ingredients which are derived from renewable resources that cannot be exhausted and consequently do not affect the next generation (Muthu et al. 2010). To achieve sustainable development, designers need to be aware of environmental impacts and incorporate environmental awareness into the design (Adams and Frost 2008). Product sustainability is the easiest aspect to alter for an apparel brand, as this is where a company has the most and direct control through design and product development. Transforming product sustainability may be achieved via various aspects such as fiber/textile selection, processing methods, use behaviors, and reuse/recycle strategies. Material selection is often the first step that designers and product developers will take in reducing the environmental impact of the garment. Environmentally preferred fibers/textiles can significantly reduce the environmental impact and increase the resourcefulness of an apparel product throughout the garment life cycle without changes to design practice or product development processes (Allwood et al. 2006).

2.2 Environmental impact of textiles and clothing products

The apparel industry has many negative environmental and social impacts that are complex and occur at different stages of the apparel life cycle. The development of fast fashion has amplified the impacts due to the increased volume of apparel produced and sold at low prices (Laursen et al. 2007). The apparel has a long and complicated production chain consisting of many phases including resource production and extraction, fiber and yarn manufacturing, textile manufacturing, apparel assembly, packaging, transportation and distribution, consumer use, recycling, and ultimate disposal (Islam and Khan 2014). The environmental impacts of apparel are varied across the phases, difficult to assess for individual garments, and are subject to the type of raw material used, dyeing and laundering. The major environmental impacts associated with the production and use of apparel throughout its life cycle include wastewater emissions from dyeing, finishing, and washing processes, increase in pollution, solid waste production, and significant depletion of resources from consumption of water, fossil fuels, and raw materials (Islam et al. 2014). Energy is used for laundering, transportation, operations of machines for various processes, the production of primary materials, especially man-made fibers such as polyester (a petroleum-based product), and yarn manufacturing of natural fibers such as cotton (Laitala and Boks 2012). Conventional cotton production has high water consumption and employs the use of toxic chemicals that may harm human health and the environment (Allwood et al. 2006). Chemicals are also released in wastewater from processes such as pre-treatments, dyeing, finishing, and laundry. These chemicals are disruptive to both the environment and aquatic-based life (Ruskino 2007). Solid waste is produced during natural fiber yarn, textile, and apparel manufacturing and disposal of apparel products at the end of their life. There are significant issues with clothing waste as the majority of clothing and textile wastes ends up in landfills as opposed to being recycled or reused (Goworek 2011). However, environmental impacts cannot be addressed once the product is introduced into the market and more attention should be focused on eco-environmentally friendly product design. Improvements in product design can be made by improving the flow of life cycle information and eco-design guidelines, integrating environmental considerations into the manufacturing processes, and involving relevant stakeholders to review the approach (Steinberger et al. 2009). Education on the need for eco-friendly products and the use of case studies and examples would help to strengthen the design and manufacture of environmentally sustainable products. European Commission-Health and Consumers, Scientific Committees Opinion (1999).

2.3 Impact of materials selected to eco-sustainability

Of all the life stages, the apparel industry has concentrated its environmentally responsible innovation in developing eco-friendly materials, for some understandable reasons. Material choices incorporate the environmental impacts of the production of those materials into the apparel product's life cycle and affect how the garments can be processed, cared for, and disposed (Lewis and Chem 2006). The environmental concerns for materials include energy consumption, use of toxic chemicals, ethics, and sustainability of resource consumption (Muthu et al. 2009). In order to address these material environmental concerns, eco-friendly apparel designers must strive to minimize non-renewable resource and energy use, as well as waste, while utilizing materials that are nontoxic and ethically produced. These materials should also provide valuable characteristics to the consumer, require low-impact care, and be effectively disposed through reuse, recycling, energy reclamation, or composting (Goan 1996). The apparel designer's primary material is of course fabric that can be composed of various fiber types that have unique properties and production. The two principal fiber types are natural fibers from plant cellulose and animal proteins and manufactured fibers derived from natural and synthetic polymers. Fibers are either cultivated, manufactured, or a combination of both. Each fiber source and production method come with its own set of environmental concerns, yet all can be improved to various degrees of reduced environmental harm by either employing best practices in current systems, implementing new, more effective technologies, or replacing current materials and practices with eco-friendly alternatives.

2.4 Impact of manufacturing processes to environmental sustainability

Clothing production is a multi-step procedure that often includes spinning, weaving, knitting, pre-treating, dyeing and/or printing, finishing, and makeup (cut-sew-trim). During these stages, the apparel product is subjected to various chemical treatments, many of which have been highly toxic and non-degrading, but now are sometimes replaced by more nontoxic and biodegradable counterparts (Kadolph 2010). Some fiber processing standard exists that prohibit the use of hazardous chemicals, including the Organic Trade Association's (OTA), American Organic Standards for Fiber Processing (Murray and Coody 2003), and the Global Organic Textile Standard (GOTS). Each production stage demands energy and often water to apply treatments, which then require more energy to dry the product after treatment. And lastly, the working conditions at each stage can pose health and safety risks due to exposure to toxic chemicals and air and noise pollution. Eco-friendly apparel production uses resources and energy effectively, only uses nontoxic cleaners and applications. Water is used during many fabric production stages for cleaning and application of dyes and finishes, resulting in large amounts of problematic wastewater (Nese and Ismail 2007). This wastewater can contain washed away dirt and wax from natural fibers, cleaning chemicals, auxiliaries, etc. Most wastewater is dumped rather than treated due to the expense, causing disruptions in the local environment and contaminating the fresh water supply. Dyeing can be done at almost any stage, from the fiber to the constructed product, and may require whitening pre-treatment (Ruskino 2007). Dyes and printing inks require a pigment and a fixative, or mordant, both of which can utilize toxic chemicals and heavy metals. Azo group chemicals, which release carcinogenic arylamines, are widely used in synthetic dyes and pigments. Dyes are difficult to remove from wastewater and leave it colored, preventing wastewater from supporting plant life by inhibiting aquatic plants' ability to photosynthesize. Eco-friendly alternative color options include low-impact and natural dyes that use nontoxic mordants, water-based inks, and color-grown cotton and wool fibers, although the color-grown palates are often muted and limited (Challa 2014). Finishing occurs after the fabric is made and often implies the application of chemicals to the fabric to give it enhanced properties, either esthetic or functional. This process often requires water for the application and to wash out the excess finish. Many finishing processes have been redesigned to use less water or foam substitute, although they still need to be improved to reduce water use and increase wastewater processing. Toxic chemical finishes pose human health risks while their wastewater disrupts ecosystems, yet more finishes are being replaced with less toxic, biodegradable alternatives. Certain finishes can also require the garment to be "dry-clean only," which uses cleaning solvents with toxic chlorine chemicals. Conversely, finishes can make traditionally dry clean only fibers, like wool and silk, washable. Some, such washable finishes are toxic chemicals while other uses environmentally safe, natural enzymes (Chen and Burns 2006). Make-up, also known as cut-make-trim, involves cutting the garment pattern out of fabric, sewing it together, and then adding any trim or embellishments. By eliminating waste in the overall process, the product's manufacturing sustainability can be increased remarkably (Islam et al. 2014). Wastages from cutting can vary between 6 and 25 %, depending on the complexity of the pattern, although this cutting waste is often recycled into low-quality textile products (Laursen et al.2007). The apparel industry still relies heavily on human labor as no machinery has been able to reproduce the agility and intricacy of a human producer.

2.5 Introduction to the Sustainable Apparel Coalition and Higg Index tool

The world's textile and apparel industry cover various areas that include the manufacturing, marketing, and retailing of textiles and garment products. This industry has been considered as an approach for industrialization, economic progress, and national development. According to the World Trade Organization, China has been leading the world with regard to export in the field of textiles and apparel, followed by the European Union and India (Wolf et al. 2011). Despite many recent recessions, strategic policies taken in this sector have saved the industry from various problems. Every industry should concentrate on keeping stock levels low, as well as on being flexible and in fine tune with the consumer's needs and wants, emphasizing lean management and strong supply chain networks. Ecological friendliness is the main motto and consumers have been very much aware of its impact, seeking out products that complied with ecological standards (Speer 2005). However, the Sustainable Apparel Coalition (SAC), which helps industries to rate their products with a numerical sustainability, scores in order to provide data to customers with regard to the extent the manufacturer has contributed to the conservation of the environment. SAC (2013) The Higg Index is one such tool for the assessment of product's sustainability, which raises a manufacturer's consciousness of the design, choice of raw materials, manufacturing processes, and finishing, packaging, and distribution through the use and recycling of the product (Martin 2013; Reichard 2013). The SAC is a multi-stakeholder engagement, formed in 2011, by a group of global apparel and footwear companies and non-profit organizations representing nearly one third of the global market share for apparel and footwear. They seek to build a common approach for measuring and evaluating apparel and footwear product sustainability performance (Ulibarri 2011). It aims to develop common measurements and a common environmental understanding of a product impact across the industry by building on the Outdoor Industry Association (OIA)'s, Eco Index™, and Nike's Environmental Design tool (Nike Inc. 2012c). The Eco Index™ is a standardized tool for measuring the environmental impacts of outdoor products such as clothing and tents and evaluates the impacts in six key areas of a product's life cycle, namely materials, packaging, product manufacturing and assembly, transport and distribution, use of services, and end of life (Nike Inc. 2012c). Measuring environmental performances of apparel products will spotlight priorities for action and opportunities for technological innovation. SAC (2013c) The Higg Index is an "indicator based tool for the apparel that enables companies to evaluate material types, products, facilities and processes based on a range of environmental product design choices" (The Higg Index 2012). The Higg Index is a set of assessment tools that are used to evaluate the environmental impact of apparel and footwear products. The index was initially released on 26 June 2012 as the Higg Index 1.0, and it has been used by many organizations—both SAC members and nonmembers. The Higg Index 1.0 used a Microsoft Excel interface and worked on qualitative indicators for assessment. The sustainability topics were related to the environment, and the product category was apparel.

Approximately 44 materials were included in the material assessment data, and the material sustainability index used basic indicator questions on the environment. The Higg Index version 2.0 is the updated version of 1.0 and was released on 11 December 2013. It is based on tools such as the Eco Index, Nike's Apparel Environmental Design Tool, Global Social Compliance Program reference tools, and social/labor practice tools. After a pilot testing period and use of the second version by many organizations for over 14 months, the Higg Index 2.0 was introduced SAC (2013b). This tool helps to standardize the methodology for measurement and evaluation of the environmental performance of the apparel products along the supply chain in three levels—namely the brand, product, and facility levels. The scope of the Higg Index 2.0 is to assess the environmental and social/labor performance of apparel and footwear products. It is based on life cycle analysis spanning the entire lifecycle of apparel products, encompassing raw materials, manufacturing, packaging, transportation, use, and end of life. The Higg Index 2.0 is a tool that educates small and large companies to recognize challenges and sustain improvement. The self-assessment tool helps scientific learning by means of identifying the vital aspects of environmental sustainability and opportunities that will provide improvement. This index is the basis for future operations and efforts to ensure sustainability and also the starting point for the commitment, learning, and collaboration among stakeholders. The scope of the desired outcomes of the SAC Higg Index tool includes improvement to reduce water use and improve quality, reduce energy and emissions, minimize waste, reduce chemicals and toxicity, and increase transparency of social and labor issues (Ruskino 2007). The tool evaluates material type, products, facilities, and processes. By utilizing practice based, qualitative binary yes/no questions, assessment can be made as to the sustainability performance of the product and drive behavior for improvement. The Higg Index RDM-Beta is a prototype which aims at educating and providing quick directional guidance to apparel designers during the product creation process about the potential environmental impacts of their design solutions (SAC 2013a). It is to engage designers in the Higg Index's product life cycle thinking and how we directionally assess materials sustainability through the MSI. Through self-assessment, organizations can better understand the environmental impacts that occur throughout the life cycle and the effects of design choices such as material type SAC (2012i). However, there are many apparel brands belonging to the Higg Index such as: H&M, Gap Inc., Nike, Adidas, Puma, Patagonia, Mountain Equipment Co-op (MEC), Levi's, Hanesbrand, Marks and Spencer, Esprit, Columbia, Timberland, Loomstate, S. Oliver, and many such bold face names. These brands are widely using this Higg Index tool to measure environmental performance of their apparel products towards better sustainability of their products (Reuben 2013).

3. METHODS

To carry out this study, selected knitted men's T-shirt products of 100 % cotton single jersey fabric was taken into consideration as a case study. Both foreign and local branded knitted men's "M" sized short sleeve T-shirt products from some renowned brand were selected on a random basis to measure their environmental performance. The brand B-, C-, and D-named T-shirts were collected from an apparel industry named Viyellatex Group Ltd. T-shirts from brand E and F were collected from Beximco fashion wear and Aarong. BUTex-Innovation T-shirt was developed by the researcher as eco-friendly product which is indicated as brand A in this study. However, environmental sustainability of mentioned branded T-shirt was evaluated based on mainly two aspects only. One is to perform selected eco-friendly tests and another is an evaluation of the products by SAC Higg Index tools. Although, some other aspects are related to a fully environmentally sustainable apparel product.

3.1 Product environmental aspect test procedures

To evaluate environmental sustainability of base materials, an initiative was taken to carry out the eco-friendly test of product base body fabric that included formaldehyde, pH value, analysis of amines in azo dyestuff content, and alkylphenol ethoxylate (APEO) content test. These entire eco-friendly tests were performed as per standard test procedure. The standard test procedure followed for various test were the following: for formaldehydes - International Standard ISO 14184–1: 2011 (for textile and non-woven), for pH-ISO 3071:2006 extraction with potassium chloride (for textiles), for analysis of amines in azo dyestuffs - European Standard EN 14362–3: 2012, (For textile) and for alkylphenol ethoxylate (APEO) content - methanol extraction and analysis of by liquid chromatograph mass spectrometer (LC-MS). If the test results are out of range expected as per standard, then that products is termed as harmful to environmental sustainability in that aspect.

3.2 T-shirt materials and manufacturing environmental aspect evaluation by SAC Higg Index tool

All the six T-shirt products were evaluated by the SAC Higg Index tool that provides a value that represents how environmentally responsible an apparel item is in terms of sustainability. For this study, the tool Higg Index was used to obtain product scores for the abovementioned T-shirts. The final score was obtained by answering questions relating to the product materials and manufacturing environmental sustainability issues. The data used during this process include information readily available to the manufacturers, suppliers, merchandisers including hang tags, care labels, and product technical package descriptions, and throughout the LCA phase of the T-shirt. The product technical package detail information was the important source for various Higg Index-required information towards eco-sustainability evaluation. Higg Index requires that all materials present be entered into a material input table along with eco-manufacturing issues. This includes a description of all materials used in addition to information obtained from follow-up questions. By utilizing practice-based, qualitative binary yes/no questions, assessments were made as the sustainability performance of the product and drive behavior for improvement. Higg Index is a Microsoft Office Excel and web-based tool that has predetermined standard values based on a global survey for specific materials that are used in determining a score. Various links between inputs and outputs considered were based on Higg Index tools as the tool has predetermined standard values based on global survey and mathematical model for specific materials and manufacturing process issues. The higher the Higg Index score, the better the environmental performance of an apparel products.

4. RESULTS AND DISCUSSION

In this study, the all five selected foreign and local branded knitted T-shirts along with newly developed one were assessed using the Higg Index 1.0 tools to obtain a value that represents a product's environmental sustainability especially in material and manufacturing issues. A higher score implies a more sustainable product, and a lower score implies less sustainability. Upon completion of the excel-based Higg Index tool, the brand A, B, C, D, E, and F T-shirts received different scores on a scale of 1–100. The superior score of the BUTex-Innovation (brand A) and the foreign branded products, especially the brand B-shirt, was due to the organic material content, eco-friendly raw materials and process used product end of life options and confirmed traceability. On the other hand, the local branded T-shirt gained comparatively lower score in most of the areas due to not using eco-friendly raw materials, manufacturing processes, high-impact finishes, improper or insufficient care and repair instructions, and traceability issues. This also assisted to find out various gaps regarding apparel product materials and manufacturing issues and to take steps to improvement of superior clothing product in terms of sustainability.

4.1 Results for environmental evaluation test of T-shirt product

Environmental issues are playing an increasingly important role in consumer products. Textiles and clothing are also subjected to stringent examination, particularly with regard to the chemicals used in their manufacture as they are often in direct contact with the skin. The consumer is confronted with scare stories of health risks and pollution allegedly originating from modern textiles treated with "poisonous, allergenic, or carcinogenic dyes and finishing auxiliaries". Such gross misrepresentation is completely untrue for textile production in countries, where extensive environmental and chemical pollution laws exist (GB 18401 2010). Nevertheless, it cannot be ruled out that some textiles of uncertain origin traded on international markets may have been processed inadequately and may contain residues of untested, possibly hazardous, auxiliaries and dyes. In light of this, the responsible partners in the textile industry believe that it is important to reassure consumers about textiles in human ecological terms, i.e., to reassure the public that properly processed textile products present no health risks (GB 18401, 2010). However, various eco-friendly tests performed for the above selected T-shirt product item showed significant difference.

4.1.1 Formaldehyde test

Formaldehyde has allergenic and sensitizing effects, even at low concentrations. Levels of 300-mg/kg formaldehyde on a garment can trigger allergic effects in sensitized individuals. Formaldehyde has an irritant effect on the respiratory tract and mucous membranes in concentrations above 0.5 mg/m^3 air. Test method followed for formaldehyde is the International Standard ISO 14184–1: 2011 (for textile and non-woven). The T-shirts of A, B, C, and D brands showed better results in formaldehyde detection with value 11.42, 20.47, 17.82, 32.45 ppm, respectively. But a local branded T-shirt of E and F brand showed poor results with higher value, i.e., 46.23 and 62.35 ppm, respectively. Results are shown in Table 1.

Table 1. Results of product eco-friendliness test for various selected brand

Branded T-shirt	Results of formaldehyde test (in ppm). (Detection limit: 16 ppm) (Maximum limit: 75 ppm)	pH value test (Maximum limit: 4.0–7.5 ppm)	Results of analysis of amines in azo dyestuffs (in ppm) (Detection limit: 10 ppm)	Results of alkylphenol ethoxylate (APEO) content test in ppm (Detection limit: 30 ppm) (Maximum limit: 100 ppm)
Brand A	11.42	6.4	Not detected	Not detected
Brand B	20.47	5.7	Not detected	Not detected
Brand C	17.82	6.2	Not detected	Not detected
Brand D	32.45	5.5	13.5	Not detected
Brand E	46.23	7.4	31.7	47.6
Brand F	62.35	7.5	43.2	64.3

Various test methods—formaldehyde test method: International Standard ISO 14184–1: 2011 (for textile and non-woven); pH value test method: ISO 3071:2006 extraction with potassium chloride (for textiles); analysis of amines in azo dyestuff test method: European Standard EN 14362–3: 2012. (for textile); alkylphenol ethoxylate (APEO) content test method: methanol extraction and analysis of by liquid chromatograph mass spectrometer (LC-MS) (for textile)

4.1.2 pH value test

The pH value of an aqueous extract of the textile article is determined according to ISO 3071. The required pH range corresponds approximately to the natural pH value of human skin (approximately 5.5). This avoids skin irritation due to the release of acidic or alkaline substances from damp textiles. The pH maximum limit for safe textile materials is 4.0–7.5. However, the tested pH value results for various branded T-shirts were 6.4, 5.7, 6.2, 5.5, 7.4, and 7.5 for brand A, B, C, D, E, and F T-shirts, respectively (Table 1). Also, noticed that the pH value for local brand E and F T-shirts was above 7.0.

4.1.3 Analysis of amines in azo dyestuffs

Azo dyes are compounds characterized by their vivid colors and provide excellent coloring properties. They are important and widely used as coloring agents in the textile and leather industries. The risk in the use of azo dyes arises mainly from the breakdown products that can be created in vivo by reductive cleavage of the azo group into aromatic amines. Due to the toxicity, carcinogenicity, and potential mutagenicity of thus formed aromatic amines, the use of certain azo dyes as textile colorants and the exposure of consumers using the textile colored with azo compounds cause a serious health concern. The two main routes of consumer exposure are the skin absorption of the azo compounds from the dyed clothes worn and potential oral ingestion, mainly referring to the sacking of textiles by babies and young children. In this study, test method followed by analysis of amines in azo dyestuffs is the European Standard EN 14362–3: 2012 (for textile). From Table 1, it is seen that the A, B, and C branded T-shirts showed better results i.e., no detection of such substances, but in D, E, and F branded T-shirts, these substances were detected with values 13.5, 31.7, and 43.2 ppm, respectively. Use of eco-friendly raw materials and third party certified dyes and chemicals may add to better results for A, B, and C branded T-shirts.

4.1.4 Product test for alkylphenol ethoxylate (APEO) content

Analyses for APEO have been a fast growing issue of textile material testing. Surprisingly to the producers, APEOs have been detected in rather high amounts in textiles. The most important APEOs or alkylphenol ethoxylates for the textile industry are NPEOs (nonylphenol ethoxylates) and OPEOs (octylphenol ethoxylates) due to their detergent properties. About 90 % of the produced APEOs are in fact NPEOs. The compounds are used in detergents, cleaning agents, or chemicals used for textile or leather production. Toxicological characteristics are their hormone-disruptive properties and the fact that they are toxic to aquatic organisms (Ljungberg 2007). Furthermore, APEOs are very persistent and difficult to be degradable in nature. These are all problem areas for wastewater treatment and discharging of wastewater into surface water. In this study, standard test method followed for textile is methanol extraction and analysis of by liquid chromatograph mass spectrometer (LC-MS). No detection of APEO was found for A, B, C, and D branded T-shirts due eco-friendly dyes and chemicals used in manufacturing processes. Alternatively, a value of 47.6 and 67.3 ppm was detected for E and F branded T-shirt test which might be due to environmentally harmful dyes and chemicals and raw material selection in manufacturing stages. Table 1 presents the result details for APEO test.

4.2 Test results for various branded T-shirts

4.2.1 Higg Index evaluation results for materials

The materials selected for the knitted T-shirt for both foreign and local branded products were entered into the material input table in the Higg Index tool. The values obtained by various products are shown in Table 2. The developed men's T-shirt A and foreign brand B, C, and D obtained higher score. This was due to the eco-friendly raw materials used, i.e., brand A and B T-shirt included 100 % organic cotton fiber. During processing of fabric, BUTex-Innovation T-shirt, B-, and other foreign branded T-shirt used eco-friendly, third party verified and environmentally low-impact dyes and chemicals. But in the case of local branded E and F T-shirt, the Higg Index score was lower due to environmentally hazardous conventional raw materials, processing, poor quality dyes used, no third party verification and such types, and many other issues. For better environmental sustainability, these issues must be addressed for local branded apparel products. Obtained score out of 100 points by A, B, C, D, E, and F branded T-shirts are 63.3, 58.8, 51.6, 46.6, 39.6, and 30.6, respectively. Major variation issues in scoring of branded T-shirt by Higg Index are shown in Tables 3 and 4.

The material sustainability index (MSI) also examines chemistry, energy, and GHG intensity, water- and land-use intensity, and physical waste across the entire cradle-to-gate life cycle of a material. Land-use intensity focuses solely on the origin of a raw material in phase 1 and does not consider land use elsewhere in the material life cycle. The MSI compiles and scores data for 100 % cotton knit (For A, B, C, D, E, and F branded products) fabric based on standard global survey data compiled and established by 13 individual indicators within the various impact areas (MSI 2013).

4.2.2 Higg Index evaluation results for manufacturing stages

The manufacturing stage is obviously an important area for better product sustainability. Some major issues have huge impact on product environmental sustainability as per Higg Index tool. The developed men's T-shirt, B-, and other foreign branded T-shirt used low-impact garment finish, i.e., process combination enzyme wash or basic rinse that scored higher. But local branded F and E T-shirts used high-impact garment finish, namely traditional bleach and acid chemical with stone wash, and scored lower. Besides, marker efficiency has great impact on product sustainability. The higher the marker efficiency, the better the product sustainability; and lower marker efficiency indicates its direct impact on supply chain, including more wastage of material, fabrics, dyes, chemicals, process loss, land fill, etc. As developed BUTex-Innovation brand A, B, C, and D branded T-shirts had more than 91 % marker efficiency, they scored better. But E and F branded T-shirts had marker efficiency of 82 and 85 % and scored lower in the Higg Index tool. Similarly, a single colored, screen-printed logo with small area received a higher score for foreign branded T-shirts and these issues vice versa scored lower for local branded T-shirts. The total points for manufacturing stages by various branded T-shirts are 60, 72, 65, 63, 49, and 27 for A, B, C, D, E, and F respectively which is shown in Table 5.

Table 2. Total materials sustainability score for A and B branded T-shirts

Materials eco-sustainability scoring issues	Resultant answer (For A and B brand)	Obtained materials Score points (A)	Obtained materials Score points (B)
Materials name	Fabric (100 % organic cotton knit fabric	28.2 (out of 50)	28.2 (out of 50)
Certified recycled content %	0 %	0.0	0.0
Virgin, certified organic content %	100 %	10.0	10.0
Other source certification or verified chain of custody with name	Yes, GOTS, BCI certified	7.0	7.0
Coating/laminate finish? (If yes, specify type)	No	4.0	4.0
Reduced water use in dyeing/coloring? (If yes, specify reduced water process/technology)	Yes, used process or technology that reduces water consumption (and reduction is documented), eco-friendly dyes and chemicals used, enzymatic scoring used	5.0	5.0
Chemical impact reduction ("green chemistry") considered?	Yes	4.5	4.5
Green chemistry third party verification?	"Yes" OEKO-Tex Standard-100 (class-II) verified) but "No" for BUTex-Innovation T-shirt	0.0	4.5
		58.8 (out of 100)	Total = 3.3 (out of 100)

Table 3. Total materials sustainability score for C and D branded T-shirts

Materials eco-sustainability scoring issues	Resultant answer		Obtained materials Score points (C)	Obtained materials Score points (D)
	C branded T-shirt	D branded T-shirt		
Materials name	Fabric (100 % cotton knit fabric	Fabric (100 % cotton knit fabric	26.6 (out of 50)	26.6 (out of 50)
Recycled content %	0 %	0 %	0.0	0.0
Virgin, certified organic content %	0 %	0 %	0.0	0.0
Other source certification or verified chain of custody with	Yes, GOTS, BCI certified	Yes, BCI certified	7.0	7.0
Coating/laminate finish? (If yes, specify type)	No	No	4.0	4.0
Reduced water use in dyeing/coloring? (If yes, specify reduced water process/technology)	Yes, used process or technology that reduces water consumption (eco-friendly dyes and chemicals, enzymatic scoring used)	No, used traditional dyeing/ coloring process or technology	5.0	0.0
Chemical impact reduction ("green chemistry")?	Yes	Yes	4.5	4.5
Green chemistry third party verification?	Yes, OEKO-Tex Standard-100 (class-II) verified)	Yes, OEKO-Tex Standard-100 (class-II) verified)	4.5	4.5
			Total = 51.6 (out of 100)	Total = 46.6 (out of 100)

Table 4. Total materials sustainability score for E and F branded T-shirts

Materials eco-sustainability scoring issues	Resultant answer		Obtained materials Score points (E)	Obtained materials Score points (F)
	E branded T-shirt	F branded T-shirt		
Materials name	Fabric (100 % cotton knit fabric	Fabric (100 % cotton knit fabric	26.6 (out of 50)	26.6 (out of 50)
Recycled content %	0 %	0 %	0.0	0.0
Virgin, certified organic content %	0 %	0 %	0.0	0.0
Other source certification or verified chain custody and name	No	No	0.0	0.0
Coating/laminate finish? (specify type)	No	No	4.0	4.0
Reduced water use in dyeing/coloring? (If yes, specify reduced water process/technology)	No, used traditional dyeing/coloring process or technology	No, used traditional dyeing/coloring process or technology	0.0	0.0
Chemical impact reduction ("green chemistry")?	Yes	No	4.5	0.0
Green chemistry third party verification?	Yes, OEKO-Tex Standard-100 (class-II) verified)	No	4.5	0.0
			Total = 39.6 (out of 100)	Total = 30.6 (out of 100)

Table 5. Higg Index evaluation on manufacturing eco-sustainability of various branded T-shirts

Manufacturing eco-sustainability of various branded T-shirts	Answers of questionnaires						Score Point Obtained (out of 100)					
	Brand A	Brand B	Brand C	Brand D	Brand E	Brand F	Brand A	Brand B	Brand C	Brand D	Brand E	Brand F
Marker/material efficiency %	91	93	95	93	85	82	41(out of 50)	43	45	43	30	24
Garment finish	Basic rinse	Enzyme wash	Basic softener	Basic softener	Basic softener	Acid chemical wash	10 (out of 30)	20	10	10	10	0
Applied graphics and logo	10 cm^2 or less	10 cm^2 or less	10 cm^2 or less	10 cm^2 or less	10–100 cm^2	100–200 cm^2	6.0 (out of 10)	6.0	6.0	6.0	6.0	2.0
Screen printing size and color	10–100 cm^2 and 3 color	10–100 cm^2 and 3 color	10 cm^2 or less and 2 color	10 cm^2 or less and 2 color	10–100 cm^2 and 3 color	100–200 cm^2 and 5 color	3.0 (out of 10)	3.0	4.0	4.0	3.0	1.0
Total							60	72	65	63	49	27

Integrating issues for environmental sustainability improvement of apparel products

The researcher has identified many technical requirements and issues for apparel product sustainability improvement based on the Higg Index assessment tool.

4.3 Sourcing fair production and practice transparency:

Sourcing fair production can be performed by selecting manufacturers that are known to operate in an eco-socially responsible manner. For the technical requirement of practicing transparency, the researcher suggests a hang tag including maximum detail information to the product that includes the name and contact information of a hypothetical manufacturer. This would allow a customer to verify environmentally and socially responsible methods practiced when the product was produced.

4.4 Consider waste-reducing design and clean manufacturing policy:

To reduce waste, the researcher suggested using eco-friendly dyeing methods such as reduced water dyeing or waterless dyeing method for the fabric processing. An on-product label with instructions to recycle, reuse of the product, thus reducing waste, should be added to the label. Clean production can be achieved by selecting manufacturers that make a commitment to the use of alternative or renewable energies. Maximum possible marker efficiency should be ensured during product development and manufacturing. Product printing, washing, and other finishing materials and processes should be selected focusing eco-sustainability priorities.

4.5 Raw materials:

Selecting the eco-friendly raw materials is an important aspect for minimizing overall environmental impact throughout the product life cycles. Organic textile farming uses up to 60 % less water than conventional farming methods and were grown without cancer-producing pesticide and insecticide. Organic clothing is made from organic fibers grown without both exposures to toxins and irreversible environmental damage. Key factors of sustainable clothing are the fiber source and renewability without the use of agro-chemicals, hormones, and pesticides, along the entire manufacturing process from raw fiber to textile.

4.6 Consider natural and low-impact dyes and chemicals:

The designers, manufacturers, and merchandisers should introduce natural and environmentally low impact dyes in material processing for better product sustainability. Dyes and chemicals used for a product in wet processing, finishing, washing, and printing process should address both environmental concerns and economic realities with consumer demand for greener products.

5. CONCLUSIONS

The textile and apparel industries are notorious for their excessive contribution of waste and pollution to our environment. The unsustainable practice trends of poor designing, raw material selection, processing, manufacturing, and disposing of apparel products are the major contributing factors to this problem. The purpose of this study was to evaluate the material and manufacturing environmental sustainability of various branded T-shirts and find out various sustainable issues for greener apparel products. Based on eco-friendly test and Higg Index environmental assessment standard tool, the findings of this study could be helpful for many organizations and customers who attempt to produce and consume sustainable textiles and clothing products. Both the local and foreign branded products should be produced by strictly maintaining sustainability issues. Hence, this study is a little approach towards better sustainable practices and eventually would drive the textile and clothing designer, manufacturers, merchandisers, and consumers to introduce various sustainable issues in their works. Hence, this study leads to growing awareness and hopefully would direct them towards sustainable practices in the textile and clothing business for a greener world.

COMPETING INTERESTS

Both authors declare that they have no competing interest in relation to this study.

AUTHORS' CONTRIBUTIONS

MMI and MRK contributed to conception and design of the study. MMI carried out the experiments, analyzed the data as well as drafted the manuscript and MRK supervised the works. Both authors read and approved the final manuscript.

REFERENCES

1. Adams, CA, & Frost, GR. (2008). *Integrating sustainability reporting into management practices.*

2. Allwood, JM, Laursen, SE, de Rodriguez, CM, & Bocken, NMP. (2006). *Well dressed? The present and future sustainability of clothing and textiles in the United Kingdom.* Cambridge, UK: University of Cambridge, Institute of Manufacturing.

3. Challa L (2014) Impact of textiles and clothing industry on environment: approach towards eco- friendly textiles. http://www.fibre2fashion.com/. Accessed 28 June 2014.

4. Chen, HL, & Burns, LD. (2006). Environmental analysis of textile products. *Clothing and Textiles Research Journal, 24*(3), 248–261.

5. DEFRA (2010). Sustainable clothing action plan. https://www.gov.uk/government/uploads/system/uploads/attachment_data/file/69193/pb13206-clothing-action-plan-100216.pdf. Accessed 6 July 2014.

6. European Commission-Health and Consumers, Scientific Committees Opinion, see http://ec.europa.eu/health/scientific_committees/environmental_risks/opinions/sctee/sct_out27_en.htm, Brussels, 18 January 1999.

7. Fletcher, K. (2008). *Sustainable fashion and textiles*. London, UK: Earthscan.

8. Fletcher, K, & Grose, L. (2012). *Fashion and sustainability: design for change*. London: Laurence King Publishing.

9. GB 18401, (2010) National general safety technical codes for textile products.

10. Goan, M. (1996). *An integrated approach to environmentally conscious design and manufacturing*. Doctoral dissertation, Virginia Polytechnic Institute and State niversity, Blacksburg, Virginia.

11. Goworek, H. (2011). Social and environmental sustainability in the fashion industry: a case study of a fair trade retailer. *Social Responsibility Journal, 7*(1), 74–86.

12. Islam, MM, Khan, MMR, & Elias, K. (2014). *American Journal of Engineering Research (AJER), 03*(12), 62–68.

13. IPP (2014) Integrated product policy (IPP). http://ec.europa.eu/environment/ipp/home.htm. Accessed 25 May 2014

14. Kadolph, SJ. (2010). *Textiles. Upper Saddle River.* NJ: Person Education Inc.

15. Laitala, K, & Boks, C. (2012). Sustainable clothing design: use matters. *Journal of Design Research, 10*(1), 121–139.**View Article**

16. Laursen, S. E., Hansen, J., Knudsen, H. H., Wenzel, H., Larsen, H. F., and Kristensen, F. M. (2007). EDIPTEX: environmental assessment of textiles. Danish Environmental Protection Agency.

17. Lewis, VD, & Chem. (2006). The life of a piece of cloth: developing garments into a sustainable service system. *International Journal of Environmental, Cultural, Economic and Social Sustainability, 2*(1), 197–207.

18. Ljungberg, LY. (2007). Materials selection and design for development of sustainable products. *Materials in Engineering, 28*(2), 466–479.

19. Martin M (2013), Creating sustainable apparel value chains—a primer on industry transformation. http://www.ifc.org/. Accessed 25 Apr 2014

20. Md. Mazedul Islam and Md. Mashiur Rahman Khan, (2014). Environmental sustainability evaluation of apparel product: a case study on knitted T-shirt. Journal of Textiles. Volume: 2014, *643080*, pp: 1–6. http://dx.doi.org/10.1155/2014/643080.

21. Murray and Coody (2003), *Environmental impact of textile production.* (I. J. Ulasewicz, Ed.) New York: Fairchild Books, Inc.

22. Muthu, SS, Li, Y, Hu, JY, & Mok, PY. (2009). An exploratory eco-impact assessment of paper and plastic bags. *Journal of Fibre Bioengineering and informatics, 1*(4), 307–320.

23. Muthu, SS, Li, Y, Hu, JY, Mok, PY, Ding, X, Wang, X, & Weibang, C. (2010). Eco-impact of shopping bags: consumer attitude and government policies. *Journal of Sustainable Development, 3*, 71–83.

24. Muthu, SS, Li, Y, Hu, JY, & Mok, PY. (2012). Quantification of environmental impact and sustainability of textile fibres. *Ecological Indicator, 13*, 66–74.

25. Niinimäki, K, & Hassi, L. (2011). Emerging design strategies in sustainable production and consumption of textiles and clothing.*Journal of Cleaner Production, 19*(1), 1876–1883.

26. Nike Inc (2012c) Product design and materials. http://www.nikeresponsibility.com/report/content/chapter/materials. Accessed 2 Dec 2014.

27. Reichard R (2013) Textiles 2013: the turnaround continues. http://www.textileworld.com. Accessed 1 May 2014.

28. Reuben, A. (2013). The Higg index for sustainable apparel Environmental Performance Index. Retrieved December 12 from,http://epi.yale.edu/indicators/indicator-case-studies/reports/higg-index-.

29. Ruskino, C. (2007). Green manufacturing: an evaluation of environmentally sustainable manufacturing practices and their impact on competitive outcomes. *IEEE: Transactions on Engineering Management, 54*(3), 445–454.

30. SAC (2012i) SAC (2013) Material assessment, MSI information, MSI life cycle scope. http://www.apparelcoalition.org/msi/msi-information/msi-environmental-impact-areas.html. Accessed 8 Apr 2014.

31. SAC (2013a) Material sustainability index base material score. http://www.apparelcoalition.org/msi/. Accessed 5 May 2014.

32. SAC (2013c) Material assessment, MSI information, MSI environmental impact areas. http://www.apparelcoalition.org/msi/msi-information/msi-environmental-impact-areas.html. Accessed 2 May 2014.

33. Speer, J. (2005). Organic cotton: where, why and how. Apparel. *Journal of Sustainable Textiles Material, 46*(9), 29–34.

34. Steinberger, JK, Friot, D, Jolliet, O, & Erkman, S. (2009). A spatially explicit life cycle inventory of the global textile chain. *The International Journal of Life Cycle Assessment, 14*(5), 453–455.

35. USagain (2015). The environmental impact of cotton T-shirt. Retrieved March 15, 2015 from https://www.youtube.com/watch?v=g5rGm6veAhg.

36. The Higg Index. (2012). . Retrieved January 10, 2013 from http://www.apparelcoalition.org/higgindex.

37. Ulibarri S, (2011) sets standards for clothing and footwear. http://www.justmeans.com/blogs/sustainable-apparel-coalition-sets-standards-for-clothing-and-footwear. Accessed 20 May 2014.

38. Wolf MA, Pant R, Chomkhamsri K (2011), Towards life cycle sustainability management. http://link.springer.com/book/10.1007/978-94-007-1899-9. Accessed 1 March 2014.

8

EFFECTS OF BIOPOLISHING ON THE QUALITY OF COTTON FABRICS USING ACID AND NEUTRAL CELLULASES

Mohammad Gias Uddin

Department of Textile Engineering, Faculty of Engineering, Ahsanullah University of Science and Technology, Tejgaon I/A, Dhaka 1208, Bangladesh

ABSTRACT

The aims of the study were to analyze the treatment conditions for the use of acid and neutral cellulase enzymes and evaluate the changes in the various properties of the treated fabrics. In this study, biopolishing was carried out on bleached fabric using acid cellulase (Mega PK) and neutral cellulase (Mega L-1009D) considering three factors: concentration (0.5, 1, and 1.5 %), temperature (45, 55, and 65 °C), and time (40, 50, and 60 min). Changes in the physical properties such as weight loss, strength loss, pilling resistance, abrasion resistance (mass loss), and bending length of the biopolished fabrics were measured. Using Mega PK, weight loss and strength loss were found within 0.25 to 1.42 % and 1.7 to 24.7 %, respectively, while these values were found within 0.20 to 1.22 % and 1.5 to 18.9 %, respectively, using Mega L. On the other hand, pilling and abrasion resistance of the fabrics were improved due to biopolishing. In addition, bending length was found maximum 31.62 % lower than the untreated bleached fabric using Mega PK and 44.26 % lower using Mega L. It was also concluded that 1 %

Mega PK showed better results when applied at 55 °C for 40 min, whereas 1 % Mega L showed better results when applied at 55 °C for 50 min among all the treatment conditions.

1. BACKGROUND

Enzymes are a sustainable alternative to the harsh toxic chemicals in the textile industry (Mojsov 2014a). There is a wide range of applications and a multitude of prospects for the use of enzymes in textile processing, leading to a positive impact on the environment (Mojsov 2014b) as enzymes are readily biodegradable (Doshi and Shelke 2001). Cellulase is the most popular and versatile enzyme used in textile wet processing for biopreparation, biopolishing, and softening of cellulosic fibers (Hebeish et al. 2013; El-Sayed et al. 2010). The surface modification of these fibers to improve their outlook by removing superficial fibrils and micro fibrils conferring cooler and softer feel, brighter luminosity of colors, and more resistance to pilling propensity using cellulases is known as biopolishing (Choudhury 2006; Shah 2014; Steward 2005; Ulson de Souza et al. 2013). This is a biological process (Mojsov 2014a; Mojsov 2014b) which employs the same cellulase action on the surface of the cellulosic materials. The process can be carried out at any stage of wet processing but most conveniently performed after bleaching. In addition, this is a permanent process (Shah 2014) and it keeps the fabric in good condition after repeated washing; consequently, products become more attractive to the customer and fetch better prices.

Though the effects of cellulase hydrolysis remain as the surface phenomena, changes in many physical aspects as well as mechanical properties of fabrics take place during the processing. Improvement in the handle value is obtained on account of the changes that take place during the reaction. An important aspect of cellulase for textile application is their relatively slow kinetics which allows the modification of cellulosic fibers in a controlled manner without excessive damage (Andreaus et al. 2104). However, cellulase action should be restricted to the hydrolysis of the loose fibrils created at the surface of the fibers (Morgado et al.2000), but steam, heat, or alkali pre-treatment can facilitate cellulase penetration to the interior of the fibers and result in more severe hydrolysis (Schimper et al. 2004; Schimper et al. 2009).

Again, the removal of surface fibers or fibrils and an increase in the pilling resistance value are directly related to an increase in weight loss, which should not exceed 3–5 % (Rau et al. 2008). In addition, for knitted fabric, a weight loss of about 1–2 % is reported to be enough to release a remarkable reduction in pilling tendency (Saravanan et al. 2009). On the other hand, commercially, a strength loss of 10 % is considered acceptable due to biopolishing (Choudhury 2006). High enzyme loads and long treatment times may lead to severe fiber damage and strength loss (Andreaus et al. 2104).

The cellulase-catalyzed hydrolysis of cellulose is strongly influenced by factors such as pH, temperature, time, and agitation (Schindler and Hauser 2004). The optimal pH for a particular cellulase depends upon its origin (Schindler and Hauser 2004). Commercial cellulases for biopolishing originate from *Trichoderma reesei* and *Humicola insolens* (Azevedo et al. 2000; Heikinheimo et al. 2000). Acid cellulases are*Trichoderma*-based products which work best at pH 4.5–5.5 at a temperature of 45–55 °C (Shah 2014; Choudhury 2006). Neutral cellulases from *Humicola* are more effective at pH 6.0–7.0 and in the temperature range of 40–55 °C (Mojsov 2014a; Choudhury 2006).

The careful control of process variables such as concentration, treatment time, and temperature in an enzymatic treatment is important for analyzing the cellulase activity for the enhancement of the product. Therefore, in the present study, an attempt has been made to analyze the treatment conditions for the use of acid and neutral cellulase enzymes and evaluate the changes in properties like weight, strength, abrasion resistance, pilling resistance, and bending length of the treated fabrics.

2. METHODS

2.1 Materials

The cotton knitted fabric (single jersey) of 160 g/m^2 (yarn count 30/1 Ne) collected from Microfiber Group Limited, Narayangonj, was used throughout the research work. Acid-stable cellulase Mega PK (ECU/mL: 7000) and neutral-stable cellulase Mega L-1009D (ECU/mL: 5000) were procured from Sumcon Biotechnology, Mega Pacific Technology Inc., USA. Sodium hydroxide, sodium carbonate and Glauber's salt (Merck Specialties Private Limited), wetting agent (Primawet, DW conc.), sequestering agent (Primaquest M conc.), peroxide stabilizer, hydrogen peroxide (35 %), detergent (Primowash), and acetic acid used for the experimental work were laboratory-grade reagents.

2.2 Combined scouring-bleaching

The fabric samples were treated with a solution containing 2.5 g/L sodium hydroxide, 1 g/L Primaquest M conc., 1 g/L Primawet DW conc., and 1 g/L Primowash maintaining a liquor ratio of 1:40, at pH 10.5–11 and temperature rises up to 45 °C. Then, 35 % hydrogen peroxide (2.5 mL/L) and stabilizer (1/3 of H_2O_2) were added. The combined process was continued for 45 min at a temperature of 95 °C. Fabric samples were washed thoroughly and neutralized with 1 g/L acetic acid. Finally, the samples were washed with 1 g/L Primowash at 80 °C for 10 min.

2.3 Biopolishing

Mega PK and Mega L enzymes were used to study the effect of biopolishing on cotton fabrics. Sample size, approximately 35 cm × 35 cm, was introduced in the machine pots containing the cellulase in the appropriate buffer system and machine rpm used was 40. Three levels of cellulase concentrations (0.5, 1, and 1.5 %), temperatures (45, 55, and 65 °C), and treatment duration (40, 50, and 60 min) were used maintaining a material to liquor ratio of 1:10. Mechanical agitation was enhanced by addition of ten steel balls (each weighing 1.04 g). For acid and neutral cellulases, bath pH was maintained at 4.5–5.5 and 7, respectively. Finally, the biopolishing process was terminated by hot treatment at 85 °C for 10 min. Opticid PSD (1.5 g/L) was used as a buffering agent during Mega PK treatment.

2.4 Weight loss

Loss in weight of biopolished fabrics was measured using Eq. 1:

$$\text{Weight loss\%} = \frac{(\text{Fabric weight before biopolishing} - \text{Fabric weight after biopolishing})}{\text{Fabric weight before biopolishing}} \times 100 \quad (1)$$

2.5 Strength loss

The bursting strength of bleached and biopolished fabrics was determined using a bursting strength tester according to ISO 13938-1:1999. The strength loss of the biopolished samples was measured using Eq. 2:

$$\text{Strenght loss\%} = \frac{(\text{Fabric strength before biopolishing} - \text{Fabric strength after biopolishing})}{\text{Fabric strength before biopolishing}} \times 100 \quad (2)$$

2.6 Pilling resistance

Pilling resistance of the fabrics was determined using Martindale abrasion and a pilling tester (125, 500, and 2000 rubbing cycles) according to SN 198525:1990. The observed resistance to pilling was reported using a rating scale, showing a range pilling resistance from 1 (very severe pilling) to 5 (no pilling).

2.7 Abrasion (mass loss)

Mass loss of the fabrics was determined using Martindale abrasion and a pilling tester (100, 250, 500, 750, and 1000 cycles) according to ISO 12947-3:1998.

2.8 Bending length

The bending length of the fabrics was determined using a Shirley stiffness tester according to BS 3356:1990. A rectangular strip of fabric (6 in. × 1 in.) was mounted on a horizontal platform in such a way that it overhung like a cantilever and bends downwards. Then, the bending length was measured from the scale mark opposite a zero line engraved on the side of the platform. Each fabric was tested four times at head and tail ends of face and back sides in a wale direction, and their average was taken.

3. RESULTS AND DISCUSSION

3.1 Weight loss%

The weight loss of acid- and neutral-stable cellulase-treated cotton fabrics is shown in Table 1.

When 0.5 % Mega PK was applied on the bleached fabric for 40 and 60 min, weight loss was increased gradually with the increase of treatment temperatures as shown in Table 1. As for example, in the case of 40-min time of treatment, weight loss was found to be 0.25 % at 45 °C while it was increased to 0.27 and 0.29 % for the increase of temperature to 55 and 65 °C, respectively. But for 0.5 % PK application for 50 min, the trend was different. For sample no. 6, weight loss was found to be 0.29 % after biopolishing treatment at 65 °C which was less than 0.31 % weight loss after the treatment at 55 °C. Again, except sample no. 14, when 1 % PK was applied on fabric for the corresponding time, a similar trend of increasing weight loss with the increase of temperature was found. In addition, in the case of 1.5 % PK application, in all cases from sample nos. 19 to 27, a similar trend was found. Finally, maximum weight loss was found to be 1.42 % for sample no. 27 when 1.5 % PK was applied for 60 min at 65 °C.

Table 1. Weight loss of Mega-PK- and Mega-L-treated cotton fabrics

Sample no.	Concentration (%)	Temperature (°C)	Time (min)	Weight loss (%)	
				Mega PK	Mega L
1	0.5	45	40	0.25	0.2
2	0.5	55	40	0.27	0.24
3	0.5	65	40	0.29	0.27
4	0.5	45	50	0.26	0.31
5	0.5	55	50	0.31	0.35
6	0.5	65	50	0.29	0.38
7	0.5	45	60	0.33	0.42
8	0.5	55	60	0.35	0.45
9	0.5	65	60	0.39	0.49
10	1	45	40	0.62	0.61
11	1	55	40	0.65	0.66
12	1	65	40	0.69	0.71
13	1	45	50	0.72	0.74
14	1	55	50	0.70	0.79
15	1	65	50	0.76	0.82
16	1	45	60	0.82	0.84
17	1	55	60	0.86	0.86
18	1	65	60	0.89	0.88
19	1.5	45	40	0.98	0.95
20	1.5	55	40	1.05	0.97
21	1.5	65	40	1.12	0.99
22	1.5	45	50	1.17	1.02
23	1.5	55	50	1.21	1.06
24	1.5	65	50	1.26	1.09
25	1.5	45	60	1.32	1.13
26	1.5	55	60	1.36	1.17
27	1.5	65	60	1.42	1.22

Furthermore, weight loss was found increased with the increase of concentration of Mega PK. Using 0.5 % PK, weight loss was found within 0.25 to 0.39 %, while using 1 and 1.5 % PK, the losses were found within 0.62 to 0.89 % and 0.98 to 1.42 %, respectively. In addition, it can be said that weight loss was found increased with the increase of treatment time (except sample no. 6). In case of sample no. 6, a weight loss of 0.29 % was found when treating the fabric with 0.5 % PK at 65 °C for 50 min, which was also found in case of the 40-min treatment under similar conditions.

From Table 1, for neutral cellulase (Mega L), it was clearly found that with the increase of concentration, time, and temperature, weight loss% was increased in the case of all the biopolished samples. Maximum weight loss was found to be 1.22 % in the case of 1.5 % Mega L application on fabric for 60 min at 65 °C (sample no. 27), while minimum loss was found to be 0.2 for 0.5 % concentration of this enzyme application at 45 °C for 40 min (sample no. 1).

In general, it can be said that due to the increase of concentration, time, and temperature of cellulase application, the removal of protruding fibers from the surface of the fabric increased, and consequently under such condition, weight loss was increased and vice versa.

Mechanical agitation during the cellulase treatment in the initial stage helps cellulase adsorption and desorption processes. The removal of enzymatically loosened material from fiber surfaces leaves the fibers very clean and smooth. Due to this polishing effect, some weight loss was observed, which however did not indicate major fiber damage.

Finally, when weight loss of Mega-PK-treated samples was compared to that of Mega-L-treated samples, it was observed that the weight loss% of PK-treated samples was slightly higher in the case of sample nos. 1, 2, and 3 under lower concentration of cellulases (0.5 %). But with the increase of treatment time, weight loss decreased in the case of PK-treated samples as shown in samples 4 to 9. In samples 11 to 16, with the treatment of 1 % cellulases, weight loss% was slightly lower in PK-treated samples. In contrast, in the case of higher concentration (1.5 %) of cellulase application, weight loss was greater in all PK-treated samples than the Mega-L-treated samples (samples 19 to 27). It indicated that Mega PK was more active in action compared to Mega L as there is less chance of cellulose degradation in the neutral condition compared to the acidic condition.

3.2 Bursting strength loss

Bursting strength loss% of acid- and neutral-stable enzyme-treated samples is shown in Table 2.

When 0.5 % Mega PK was used for biopolishing of the bleached fabric, strength loss was found within 1.7 to 7.4 % under different treatment times and temperatures. In addition, strength loss was found increased with the increase of concentration of PK. As for example, when 1 % PK was used, strength loss was found within 8.2 to 15.6 %, and in the case of 1.5 % PK application, the loss was within 16.5 to 24.7 %. It was obvious in this study that the concentration of cellulase is the most important factor in influencing the strength loss.

Treatment time has also an important effect on strength loss apart from the cellulase concentration. When 0.5 % Mega PK was used for 40 min at 45 °C, strength loss was 1.7 %, while it was found to be 4.9 % only with the increase of time from 40 to 50 min and 5.8 % with the increase of time from 50 to 60 min at 45 °C. Similarly, for 1 % PK application at 45 °C, strength loss was 8.2 % for 40 min time of treatment, and this further increased to 11.5 and 14 % for time 50 and 60 min, respectively. In addition, when 1.5 % PK was applied at 45 °C for 40 min, strength loss was found to be 16.5 % and strength loss increased to 18.9 and 23.1 % for time 50 and 60 min, respectively.

But in comparison to concentration and time factor, selected temperatures within 45–65 °C did not play so much influential effect on strength loss. As for example, for 0.5 % PK application for 50 min, strength loss was 4.9 % at 45 °C; at 55 °C, it was 5 % and at 65 °C 5.8 %. Again, for 1 % PK applied for 60 min at 45, 55, and 65 °C temperatures, the strength losses were 14, 14.8, and 15.6 %, respectively, and for 1.5 % PK applied for 60 min at 45, 55, and 65 °C temperatures, the strength losses were 23.1, 23.9, and 24.7 %.

Table 2. Bursting strength loss of Mega-PK- and Mega-L-treated fabrics

Sample no.	Concentration (%)	Temperature (°C)	Time (min)	Strength loss (%)	
				Mega PK	Mega L
1	0.5	45	40	1.7	1.5
2	0.5	55	40	2.5	2.2
3	0.5	65	40	4.1	3
4	0.5	45	50	4.9	3.2
5	0.5	55	50	5	3.3
6	0.5	65	50	5.8	4.1
7	0.5	45	60	5.8	4.9
8	0.5	55	60	6.6	5.8
9	0.5	65	60	7.4	5.9
10	1	45	40	8.2	6.6
11	1	55	40	9	7.4
12	1	65	40	10.7	8.2
13	1	45	50	11.5	9.1
14	1	55	50	12.4	9.9
15	1	65	50	13.2	10.7
16	1	45	60	14	11.5
17	1	55	60	14.8	11.6
18	1	65	60	15.6	12.4
19	1.5	45	40	16.5	12.5
20	1.5	55	40	17.3	13.2
21	1.5	65	40	18.1	14
22	1.5	45	50	18.9	14.8
23	1.5	55	50	20.6	15.6
24	1.5	65	50	22.2	16.5
25	1.5	45	60	23.1	17.3
26	1.5	55	60	23.9	18.1
27	1.5	65	60	24.7	18.9

On the other hand, in the case of neutral-stable enzyme-treated samples, when 0.5 % Mega L was applied, strength loss was found within 1.5 to 5.9 % under different treatment temperatures and times. With the increase of concentration from 0.5 to 1 %, strength loss was increased and found within 6.6 to 12.4 %, and a further strength loss of 12.5 to

18.9 % was found using 1.5 % Mega L under three different temperatures and three different time combinations.

Again, for neutral cellulase enzyme, treatment time played more impact on strength loss than the treatment temperature. As for example, when 0.5 % Mega L was applied for 40 min, the strength losses were found to be 1.5, 2.2, and 3 %, respectively, at 45, 55, and 65 °C temperatures. But for 0.5 % Mega L application at 45 °C for 50 min, the strength loss increased to 3.2 %, at 55 °C for 50 min to 3.3 %, and at 65 °C for 50 min finally to 4.1 %.

For 1.5 % Mega L application to bleached fabrics at 55 °C temperature, strength losses were found to be 13.2, 15.6, and 18.1 %, respectively, for 40-, 50-, and 60-min treatment times, whereas when the temperature decreased to 45 °C for 40 min, strength loss decreased to 12.5 %, and when it increased to 65 °C for 40 min, strength loss increased to 14 %. In general, cotton fabric exhibited a loss in bursting strength due to the biodegradation of cellulose by the cellulase.

In this study, concentration had a maximum effect on strength loss followed by treatment time of 40 to 60 min. But prolonged treatment time exceeding 60 min may have significant impact on strength loss, as degradation can occur in the accessible amorphous areas of large pores and at crystalline surfaces.

The findings of strength loss found in the study indicated that the presence of increased amounts of cellulase enzyme in the biotreatment medium causes progressive interactions in a key and lock fashion between the enzyme and cotton cellulose of the fabric. In doing so, cellulose chains undergo molecular scission at 1,4 β-glucosidic linkages, thereby reducing the degree of polymerization of cotton. This, together with the loosening of the fabric structure by virtue of its loss in fabric weight, would account for the decrease in fabric strength. Moreover, biodegradation of fibers along the spiral plane of cellulose might have contributed to the initial strength loss, and when the fissures reached the lumen, further degradation in strength might have occurred rapidly.

From Table 2, it was crystal clear that more strength loss occurred in the PK-treated bleached samples than in the Mega-L-treated ones. This was because of the more effective endoglucanase (EG) activity as in the case of acid cellulases compared to the cellobiohydrolase (CBH)-rich cellulases like neutral cellulases in the formulation of the whole enzyme (Saravanan et al. 2009). EG activity leads to higher level of strength loss which could be expected from a classical endwise cleavage of the cellulose chains in the fibers. EG1 among at least six multiple EGs (Dixit and Jahan 2012) is a faster enzyme than CBH1 in the initial stages of hydrolysis, but after the cleavage of the more accessible cellulose chains on the surface of the cotton fibers, their action seems to end. In other words, EG effectively opens up more of the fiber structure to attach by CBH and β-glucanase. These two components in turn produce water-soluble fragments from the exposed area, and such synergistic action allows for rapid reduction in fiber strength.

3.3 Pilling rating

Pilling ratings of acid- and neutral-stable enzyme-treated fabrics are shown in Table 3.

The pilling resistance ratings of bleached fabric were found to be 4, 3–4, and 3 under 125, 500, and 2000 cycles of operations, respectively, in the pilling test.

Due to biopolishing using the Mega PK enzyme, the rating was improved to 4–5 for sample no. 1 and to 5 in all other 26 samples treated under different parameters for 125 cycles of operation. No change occurred in the specimen surface for samples 2 to 27 under 125 cycles of operation in the pilling test. Under 500 cycles of operation, sample no. 1 showed a rating of 4; sample nos. 2 to 7 and 9, 10, and 12 showed a rating of 4–5; and the other 17 samples (sample nos. 8, 11, and 13 to 27) showed a rating of 5. Again, under 2000 cycles of operation over the PK-treated biopolished samples, sample no. 1 showed a rating of 3–4, another 8 samples (sample nos. 2, 3, 4, 6, 7, 9, 10, 12) showed a rating of 4, 2 samples (sample nos. 5, 8) showed a rating of 4–5, and the other 16 samples (sample nos. 11, 13 to 27) showed no change in pilling testing (rating of 5).

Table 3. Pilling ratings of Mega-PK- and Mega-L-treated cotton samples

Sample no.	Concentration (%)	Temperature (°C)	Time (min)	Pilling ratings					
				Mega PK			Mega L		
				125 cycles	500 cycles	2000 cycles	125 cycles	500 cycles	2000 cycles
Bleached	–			4	3–4	3	4	3–4	3
1	0.5	45	40	4–5	4	3–4	4–5	4	3–4
2	0.5	55	40	5	4–5	4	4–5	4	3–4
3	0.5	65	40	5	4–5	4	5	4	4
4	0.5	45	50	5	4–5	4	5	4	4
5	0.5	55	50	5	4–5	4–5	5	4–5	4–5
6	0.5	65	50	5	4–5	4	5	4	4
7	0.5	45	60	5	4–5	4	5	4	4
8	0.5	55	60	5	5	4–5	5	4–5	4–5
9	0.5	65	60	5	4–5	4	5	4	4
10	1	45	40	5	4–5	4	5	4	4
11	1	55	40	5	5	5	5	4–5	4
12	1	65	40	5	4–5	4	5	4–5	4
13	1	45	50	5	5	5	5	4–5	4–5
14	1	55	50	5	5	5	5	5	4–5
15	1	65	50	5	5	5	5	4–5	4–5
16	1	45	60	5	5	5	5	4–5	4–5
17	1	55	60	5	5	5	5	4–5	4–5
18	1	65	60	5	5	5	5	4–5	4–5
19	1.5	45	40	5	5	5	5	4–5	4–5
20	1.5	55	40	5	5	5	5	4–5	4–5
21	1.5	65	40	5	5	5	5	4–5	4–5
22	1.5	45	50	5	5	5	5	4–5	4–5
23	1.5	55	50	5	5	5	5	4–5	4–5
24	1.5	65	50	5	5	5	5	4–5	4–5
25	1.5	45	60	5	5	5	5	4–5	4–5
26	1.5	55	60	5	5	5	5	4–5	4–5
27	1.5	65	60	5	5	5	5	4–5	4–5

So, the PK-treated samples numbering from 13 to 17 and sample no. 11 showed the best pilling performance. This pilling performance clearly indicated that the acid enzyme significantly improved the pilling resistance of the treated fabric samples. In fact, PK cellulase showed more activity at 1 and 1.5 % than using 0.5 %. Due to high biopolishing effect, all surface hairy fibers were removed properly. Therefore, when friction was applied on the treated fabric during the pilling test, the tendency of pill formation was significantly lower and majority caused a rating of 5, i.e., no change at the surface. A good depilling effect was also found due to mechanical agitation which accelerated the cellulase action.

On the other hand, when Mega L was used, the rating was also improved significantly from the bleached fabric. In the case of 125 cycles of operation, only sample nos. 1 and 2 showed a rating of 4–5 while the other 25 samples showed a rating of 5 as compared to a rating of 4 of the bleached sample under the same no. of cycles. In addition, under 500 cycles of operation, 8 samples showed a rating of 4 (sample nos. 1–4, 6–7, 9, 10). The remaining 21 samples showed a rating of 4–5 as compared to a rating of 3–4 of the bleached sample at 500 cycles of operation. Only sample no. 14, i.e., when a 1 % Mega-L-treated biopolished sample was applied at 55 °C for 50 min, showed a rating of 5. Under 2000 cycles of operation, sample nos. 1 and 2 showed a rating of 3–4, indicating slight to moderate surface fuzziness, while 8 samples (sample nos. 3, 4, 6, 7, 9–12) showed a rating of 4 and the remaining 17 samples showed a rating of 4–5. Rating improved in all the cases with the increase of concentration of cellulase as under 2000 cycles of pilling operation, rating was found to be 3 for the bleached sample.

Neutral cellulases originate from the fungi *H. insolens* and these are less reactive than the acid based cellulase. As Mega L cellulase was applied at pH 6–7, this required a longer treatment time to give the similar effect achieved through acid cellulases. EG and EG-rich cellulases as in the case of acid cellulases exhibit better pilling rating at lower weight losses compared to other components of cellulases. The improvements of pilling ratings with the increase of weight loss of the biopolished samples are shown in Fig. 1.

Though depilling action was significantly better in the case of Mega-L-treated samples compared to untreated bleached samples, the action was slightly lower in the case of neutral-stable cellulase-treated samples in comparison to acid-stable cellulase-treated samples, specially under 500 and 2000 cycles of operation. This is due to the more cellulase activity of the acid-stable Mega PK enzyme.

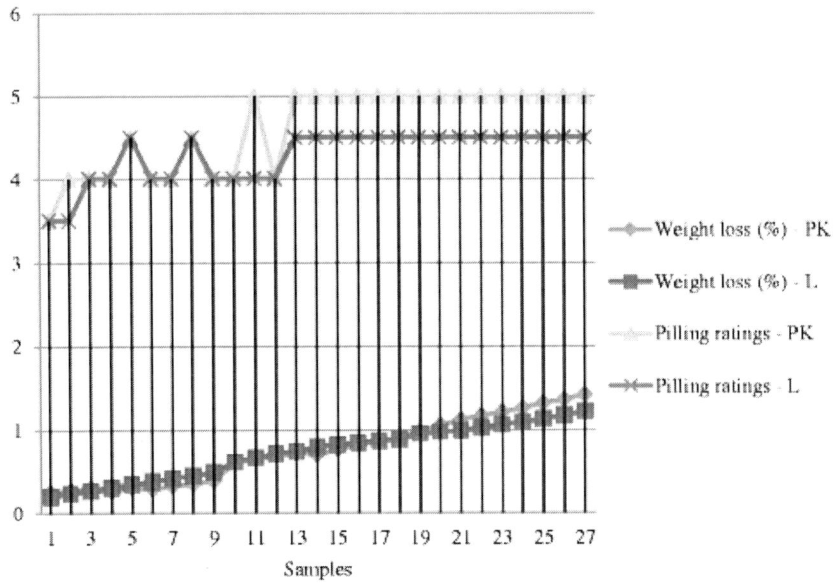

Figure 1. Weight loss% vs pilling ratings of biopolished fabrics

3.4 Abrasion resistance (mass loss)

The results of mass loss due to abrasion testing for Mega-PK- and Mega-L-treated biopolished fabrics are shown in Table 4.

From Table 4, it was obvious that mass loss increased with the increase of number of cycles during abrasion. Again, it was noticed that the amount of mass loss was decreased with the increase of concentration of PK cellulase. As for example, under 1000 cycles of abrasion testing, the mass loss of the bleached sample was found to be 1.90 mg while the maximum mass loss was found to be 1.50 mg in sample no. 1, and this gradually reduced to a minimum mass loss of 1.06 mg with the increase of concentration, time, and temperature. Only in cases of sample nos. 4 and 5 mass loss was found to be equal, i.e., 1.43 mg.

Again, under other cycles such as 100, 250, 500, and 750 cycles, mass loss was found decreased with the increase of concentration, time, and temperature with some exceptions. For 100, 250, 500, and 750 cycles, in the case of sample no. 1, maximum mass losses were found to be 0.40,

0.90, 1.06, and 1.20 mg, respectively, and minimum losses were found to be 0.02, 0.41, 0.74, and 0.96 mg, respectively, in the case of sample no. 27 (PK 1.5 %, treatment temperature 65 °C, time 60 min).

Table 4. Abrasion resistance (mass loss) of biopolished cotton fabrics

Sample no.	Concentration (%)	Temperature (°C)	Time (min)	Mass loss (mg) for Mega PK					Mass loss (mg) for Mega L				
				100 cycles	250 cycles	500 cycles	750 cycles	1000 cycles	100 cycles	250 cycles	500 cycles	750 cycles	1000 cycles
Bleached	-			0.88	1.32	1.59	1.68	1.90	0.88	1.32	1.59	1.68	1.32
1	0.5	45	40	0.40	0.90	1.06	1.20	1.50	0.60	1.10	1.42	1.53	1.72
2	0.5	55	40	0.38	0.89	1.04	1.19	1.48	0.58	1.08	1.38	1.51	1.69
3	0.5	65	40	0.37	0.88	1.03	1.18	1.46	0.57	1.09	1.35	1.49	1.65
4	0.5	45	50	0.35	0.85	1.01	1.17	1.43	0.55	1.07	1.30	1.48	1.67
5	0.5	55	50	0.33	0.82	0.99	1.16	1.43	0.52	1.05	1.28	1.48	1.64
6	0.5	65	50	0.30	0.80	0.98	1.15	1.42	0.50	1.04	1.25	1.47	1.60
7	0.5	45	60	0.32	0.76	0.96	1.15	1.40	0.48	1.02	1.22	1.45	1.62
8	0.5	55	60	0.28	0.72	0.95	1.14	1.38	0.46	1.01	1.19	1.42	1.55
9	0.5	65	60	0.25	0.70	0.94	1.12	1.35	0.44	0.99	1.18	1.40	1.53
10	1	45	40	0.21	0.67	0.92	1.10	1.37	0.41	0.97	1.19	1.37	1.53
11	1	55	40	0.18	0.66	0.90	1.09	1.32	0.40	0.96	1.20	1.38	1.52
12	1	65	40	0.15	0.65	0.89	1.08	1.28	0.40	0.95	1.17	1.36	1.48
13	1	45	50	0.12	0.62	0.87	1.10	1.30	0.37	0.93	1.15	1.35	1.48
14	1	55	50	0.14	0.60	0.86	1.09	1.27	0.35	0.91	1.16	1.33	1.47
15	1	65	50	0.10	0.61	0.85	1.08	1.26	0.32	0.90	1.18	1.32	1.45
16	1	45	60	0.08	0.61	0.84	1.07	1.24	0.30	0.87	1.14	1.30	1.45
17	1	55	60	0.07	0.58	0.83	1.06	1.21	0.28	0.86	1.13	1.29	1.42
18	1	65	60	0.06	0.55	0.81	1.05	1.20	0.25	0.84	1.10	1.27	1.40
19	1.5	45	40	0.08	0.52	0.80	1.04	1.20	0.24	0.80	1.08	1.28	1.36
20	1.5	55	40	0.07	0.50	0.80	1.04	1.18	0.23	0.78	1.09	1.26	1.32
21	1.5	65	40	0.06	0.48	0.79	1.02	1.16	0.22	0.76	1.07	1.24	1.30
22	1.5	45	50	0.05	0.47	0.78	1.01	1.14	0.20	0.74	1.05	1.22	1.29
23	1.5	55	50	0.04	0.46	0.77	1.02	1.13	0.19	0.72	1.02	1.21	1.28
24	1.5	65	50	0.04	0.45	0.76	0.99	1.12	0.18	0.70	1.01	1.20	1.27
25	1.5	45	60	0.04	0.44	0.78	0.98	1.10	0.17	0.66	0.97	1.22	1.25
26	1.5	55	60	0.03	0.42	0.76	0.97	1.08	0.15	0.64	0.95	1.20	1.22
27	1.5	65	60	0.02	0.41	0.74	0.96	1.06	0.12	0.62	0.92	1.19	1.18

Due to superficial treatment of cellulase on cellulosic fibrils and microfibrils, protruding hairy fibers are removed, as a result of which the surface of the fabric became cleaner. For this reason, resistance to abrasion was better in the case of the cellulase-treated samples than the untreated bleached sample. Again, biopolishing effect was higher in the use of a higher concentration (1.5 %) of PK cellulase; correspondingly, resistance to abrasion was better and hence mass loss was found lower.

In the case of the neutral cellulase enzyme, using 0.5 % concentration under 1000 cycles during abrasion testing, the maximum mass loss found in the case of sample no. 1 was 1.72 mg, and it decreased gradually with the increase of concentration, time, and temperature, and finally minimum loss was found to be 1.18 mg in sample no. 27. No exception was found in such trend at 1000 cycles of abrasion testing. However, except in a few cases, a similar trend was noticeable under 100-, 250-, 500-, and 750-cycle testings.

It is known that abrasion is the physical destruction of fibers resulting from frictional forces developed by mechanical repetition of lateral stresses. This causes rupture of fibers from their normal position through bending and flexing. In both acid- and neutral-stable cellulase-treated samples, a very little amount of mass loss occurred due to the smoothness of the fabric surface after biopolishing action, causing higher abrasion resistance performance.

Figure 2 shows a comparison of mass loss due to abrasion of Mega-PK- and Mega-L-treated bleached cotton samples. In Fig. 2, in all cases, the mass loss values were lower in the PK-treated samples; thus, it can be said that slightly better resistance to abrasion was found in the acid-stable cellulase-treated samples. Neutral-stable cellulases were less reactive causing less degradation to cellulose; as a consequence of which, weight loss as well as strength loss was found lower in this study. Hence, mass loss was found slightly higher during abrasion testing in the Mega-L-treated cotton samples, which showed lower resistance to abrasion.

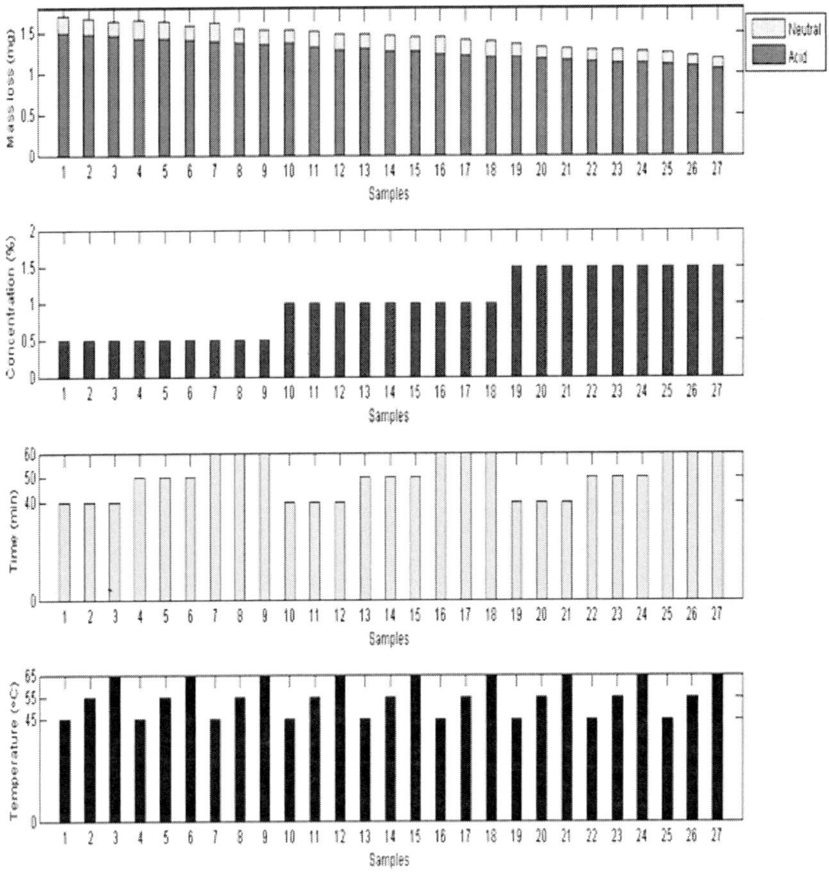

Figure 2. Mass loss (abrasion) between PK- and L-treated fabrics (1000 cycles)

3.5 Bending length

The bending length of the acid- and neutral-stable enzyme-treated samples is shown in Table 5.

The average bending length of the bleached fabric was found to be 1.48 cm in a wale direction. For Mega PK enzyme application, in 24 cases, the bending length of biopolished fabrics was found decreased than that of the bleached fabric in the wale direction. Only three samples (sample nos. 18, 22, and 25) showed a slight increase in the bending lengths which were 1.58, 1.53, and 1.61 cm, respectively. The lowest average bending length was found to be 1.01 cm in the case of sample no. 23 which was 31.62 % lower than that of the bleached sample.

Table 5. Bending length of biopolished cotton fabrics (wale direction)

Sample no.	Average bending length (cm)	
	Mega PK	Mega L
Bleached sample	1.48	1.48
1	1.23	1.4
2	1.23	1.45
3	1.1	0.98
4	1.2	1.25
5	1.4	1.63
6	1.1	1.5
7	1.2	1.33
8	1.1	1.18
9	1.1	1.03
10	1.1	1.08
11	1.1	0.85
12	1.35	0.93
13	1.32	1.3
14	1.42	0.98
15	1.24	1.46
16	1.3	1.05
17	1.36	0.9
18	1.58	1.2
19	1.4	1.11
20	1.28	1.25
21	1.35	1.13
22	1.53	1.2
23	1.01	0.83
24	1.33	1.53
25	1.61	1.15
26	1.35	1.45
27	1.18	1.55

In the application of the Mega L enzyme, the average bending length of the biopolished fabrics was decreased than that of the bleached fabric in 24 cases; only in three samples, sample nos. 5, 24, and 27, the bending lengths were 1.63, 1.53, and 1.55 cm, respectively, in the wale direction. The bending length was found lowest 0.83 cm (44.26 %) in sample no. 23. Biopolishing leads to a cleaning and depilling effect due to complete cellulase action. That is why softness was improved considerably as a result of decreased bending length.

4. CONCLUSIONS

In this study, the biopolishing effect on cotton knit fabric was investigated using two types of cellulases concerning the working parameters (enzyme concentration, treatment time, and temperature). In the case of Mega PK and Mega L, with the increase of concentration, time, and temperature, weight loss was increased in all cases, except in two treatments of PK cellulase. Fabric weight loss was found within 0.25–0.39 % using 0.5 % PK, 0.62–0.89 % using 1 % PK, and 0.98–1.42 % using 1.5 % PK. Again, it was found to be within 0.2–0.49 % using 0.5 % L, 0.61–0.88 % using 1 % L, and 0.95–1.22 % using 1.5 % L cellulase. Fabric strength loss was found to be within 1.7 to 7.4 % using 0.5 % PK, 8.2 to 15.6 % using 1 % PK, and 16.5 to 24.7 % using 1.5 % PK. Again, strength loss ranged 1.5 to 5.9 %, 6.6 to 12.4 %, and 12.5 to 18.9 % using 0.5, 1, and 1.5 % Mega L, respectively. On the other hand, pilling resistance was improved when the fabric was treated with cellulases in all cases. Abrasion resistance was improved due to biopolishing and acidic cellulase treatment showed better resistance than the neutral treatment. With the increase of concentration, time, and temperature, mass loss decreased except in a few cases. The bending length of biopolished fabrics was found lower in most cases than that of the bleached fabric. Average bending length decreased maximum 31.62 % using Mega PK and 44.26 % using Mega L. The reduction of bending length ensured the better softness of the biopolished fabrics. Finally, on the basis of the observed results, it can be suggested that Mega PK showed a better result when using 1 % concentration at 55 °C for 40 min. On the other hand, a comparatively better result was found using 1 % Mega L at 55 °C for 50 min.

COMPETING INTERESTS

The author declares that he has no competing interests.

REFERENCES

1. Andreaus, J, Olekszyszen, DN, & Silveria, MHL. (2014). Processing of cellulosic textile materials with cellulases. In JD Fontaza, M Tiboni, & A Grzybowski (Eds.), *Cellulose and other naturally occurring polymers* (pp. 11–19). ISBN 978-81-308-0543-6.

2. Azevedo, H, Bishop, D, & Cavaco-Paulo, A. (2000). Effects of agitation level on the adsorption, desorption, and activities on cotton fabric of full length and 63 core domains of EGV (*Humicola insolens*) and CenA (*Cellulomonas fimi*). *Enzyme and Microbial Technology*, 27;325–329.

3. Choudhury, AKR. (2006). *Textile preparation and dyeing* (pp. 149–150). New Delhi, India: Oxford and IBH Publishing Ltd.

4. Dixit, S, & Jahan, S. (2012). Optimization of the pre-treatment process with acid cellulase enzyme to improve physical properties of handloom cotton fabric. *International Journal of Home Economics, 5*(2), 130–139.

5. Doshi, R, & Shelke, V. (2001). Enzymes in textile industry—an environment-friendly approach. *Indian Journal of Fiber & Textile Research, 26*, 202–205.

6. El-Sayed, H, El-Gabry, L, & Kantouch, F. (2010). Effect of biocarbonization of coarse wool on its dyeability.*Indian Journal of Fiber & Textile Research, 35*(4), 330–336.

7. Hebeish, A, Kamel, MM, & Helmy HM El Hawary, NS. (2013). Science-based options for application of cellulase biotreatment and reactive dyeing to cotton fabrics. *Life Science Journal, 10*(4, 6), 3281–3289.

8. Heikinheimo, L, Buchert, J, & Miettinen-Oinonen, A. (2000). Treating denim fabrics with *Trichoderma reesei*cellulases. *Textile Research Journal, 70*(11), 969–973.

9. Mojsov, K. (2014a). Biopolishing enzymes and their applications in textiles: a review. *Tekstilna industrija, 61*(2), 20–24.

10. Mojsov, K. (2014b). Trends in bio-processing of textiles: a review. *Advanced technologies, 3*(2), 135–138.

11. Morgado, J, Cavaco-Paulo, A, & Marie-Alice, R. (2000). Enzymatic treatment of lyocell—clarification of depilling mechanisms. *Textile Research Journal., 70*(8), 696–699.

12. Rau, M, Heidemann, C, Pascoalin, AM, Filho, EXF, Camassola, M, Dillon, PC, Chagas, FD, & Andreaus, J. (2008). Application of cellulases from *Acrophial ophoranainiana* and *Penicillium echinulatum* in textile processing of cellulosic fibers. *Biocatalysts and Biotransformation., 26*(5), 383–390.

13. Saravanan, D, Vasanthi, NS, & Ramachandran, T. (2009). A review on influential behavior of biopolishing on dyeability and certain physico-mechanical properties of cotton fabrics. *Carbohydrate Polymers, 76*, 1–7.

14. Schimper, C, Keckeis, R, Ibanescu, C, Burtscher, E, Manian, AP, & Bechtold, T. (2004). Influence of steam and dry heat pretreatment on fiber properties and cellulase degradation of

cellulosic fibers. *Biocatalysts and Biotransformation., 22*(5-6), 383–389.

15. Schimper, CB, Ibanescu, C, & Bechtold, T. (2009). Effect of alkali pre-treatment on hydrolysis of regenerated cellulose fibers (part 1: viscose) by cellulases. *Cellulose., 16*(6), 1057–1068.

16. Schindler, WD, & Hauser, PJ. (2004). Chemical finishing of textiles. In *Finishing with enzymes: bio-finishes for cellulose* (pp. 181–188). Cambridge, England: Woodhead Publishing Ltd.

17. Shah, SR. (2014). Chemistry and applications of cellulase in textile wet processing. *Research Journal of Engineering Sciences, 3*(2), 1–5.

18. Steward, MA. (2005). *Biopolishing cellulosic nonwovens.* North Carolina State University: PhD Thesis.

19. Ulson de Souza, AA, Ferreira, FCP, Souza, GU, & SMA. (2013). Influence of pretreatment of cotton yarns prior to biopolishing. *Carbohydrate Polymers, 93*, 412–415.

9

THE TEXTILE INDUSTRY AND SUSTAINABLE DEVELOPMENT: A HOLT–WINTERS FORECASTING INVESTIGATION FOR THE EASTERN EUROPEAN AREA

Dorel Paraschiv, Cristiana Tudor * and Radu Petrariu

International Business and Economics Department, Bucharest University of Economic Studies, Bucharest 010374, Romania

ABSTRACT

To achieve sustainable development, massive changes towards fostering a clean and pollution-reducing industrial sector are quintessential. The textile industry has been one of the main contributors to water pollution all over the world, causing more than 20% of the registered levels of water pollution in countries like Turkey, Indonesia and China (among the G20 group of countries) and also in Romania and Bulgaria (in the Eastern European area), with even more than 44% in Macedonia. Given the controversy created by the textile industry's contribution to pollution at a global level and also the need to diminish pollution in order to promote sustainable development, this paper comparatively investigates the contribution of the textile industry to the water pollution across Central and Eastern European countries, as well as developed countries. In addition, we employ the Holt–Winters model to forecast the trend of the total emissions of organic water pollutants, as well as of the textile industry's contribution to pollution for the top polluters in Eastern Europe, *i.e.*, Poland and Romania. According to our estimates, both countries are headed towards complete elimination of pollution caused by the textile industry and, hence, toward a more sustainable industrial sector, as Greenpeace intended with the release of its 2011 reports.

KEYWORDS

pollution; sustainable industry; textile industry; Greenpeace; exponential smoothing; Holt–Winters forecasting; Poland; Romania

1. INTRODUCTION

According to the literature, sustainable development is "socially responsible economic development" that protects "the resource base and the environment for the benefit of future generations". However, in order to achieve sustainable development, massive changes in the industrial sector are needed. More specifically, to foster a significantly more sustainable industry, it is imperative to strive to achieve a clean, pollution-reducing industry.

In 2011, Greenpeace made accusations against several global brands in the fashion industry [1,2], such as Nike, Adidas, Puma, Lacoste, GAP and H & M, who use providers whose activities seriously pollute the environment in China. The report titled "Dirty Laundry", published by Greenpeace, shows that a long list of companies active in the field of fashion has commercial links with suppliers in China. The activity of these suppliers led to the discharge of chemical substances in the Yangtze and Pearl rivers, impacting the immune system function and the functioning of hormones in humans. Indeed, as shown in [3] the textile industry both uses large volumes of water (it takes approximately 200 liters of water to produce l kg of textiles) and also generates significant volumes of wastewater. Dyeing constitutes a particularly important problem in this respect, as dye houses in India and China are notorious for not only exhausting local water supplies, but for dumping untreated wastewater into local streams and rivers.

A second report issued in 2011 by the same organization [4] (*i.e.*, Dirty Laundry 2: Hung Out to Dry) attests that nonylphenol ethoxylates (NPEs) ("NPEs are a group of chemicals used in the manufacturing of textiles, in particular as surfactants and detergents. NPEs can break down to form NPs, a toxic, persistent, bioaccumulative and hormone-disrupting substance. NP is very toxic to aquatic organisms and may cause long-term adverse effects in the aquatic environment.") are present in the clothing and footwear of 15 leading brands. Ultimately, Greenpeace tried with the release of these reports to force these large companies to eliminate all releases of hazardous chemicals from their supply chains and products by the year 2020.

In 2013 the Chinese authorities have finally acknowledged the existence of so-called "cancer villages" within the country, after several years, from the emergence of information regarding the above average number of cancer occurrences in residents of some particularly polluted regions in the country. The government has also recognized that "toxic and harmful chemicals" generally prohibited in developed countries are allowed in China and "may endanger human health and the environment in the long run." (Mediafax.com).

Among the hazardous substances, nonylphenol (NP), a synthetic organic compound banned in the European textile industry, is being used extensively by the Chinese textile industry and also in the production of detergents. It is worth mentioning that although NP and NPE are currently restricted under the REACH Regulation (the Regulation on Registration, Evaluation, Authorization and restriction of Chemicals—was enforced on 1 June 2007 and was meant to improve the former legislative framework on chemicals of the European Union) for use in textile processing within the EU, this regulation does not apply in cases where there is no release to wastewater at the site. In addition, there are no restrictions concerning the NP/NPE content of finished textiles within the European Union. Furthermore, NP/NPE is not subject to international controls, so imported textiles in particular are believed to be a source of NPE in the environment, a danger to which all world countries are subject (Environment Agency, 2013).

Given the controversy created by the textile industry's contribution to pollution at a global level, but also given its proven close relationship with the overall macroeconomic growth in a developing country (see [5]), this paper comparatively investigates the contribution of the textile industry to the water pollution across Central and Eastern European countries, developed countries and China and attempts to forecast the trend of the textile industry's contribution to pollution and, hence, the road towards a more sustainable industrial sector over the next decade for the two top polluting Eastern European countries included in the analysis, *i.e.*, Poland and Romania. Previously, Domańska and Wojtylak [6] employed fuzzy time series models to forecast pollution concentrations by using real meteorological data from different areas in Poland. Furthermore, with applications on the Polish data, [7] attempted to forecast daily air pollution by using support vector machine (SVM) and wavelet decomposition. Other studies propose different time series forecasting methods to study pollution indicators. Among these, [8] proposed a parallel genetic algorithm (GA) to forecast hourly concentrations of urban air pollution in Helsinki. Ziomas *et al.* [9] proposed some analytical models relating pollutant concentrations in the Greater Athens Area with meteorological and other variables, while [10] used a Bayesian kriged Kalman filtering model. Kurt *et al.*[11] developed an online air pollution forecasting system for the Greater Istanbul Area by using neural networks, and Ibarra-Berastegi *et al.* [12] also employed neural networks to forecast five air pollution indicators in six locations in the area of Bilbao (Spain). McCollister and Wilson [13] developed two linear stochastic models to forecast the concentrations of various air pollutants in the Los Angeles area, and more recently [14] developed a model for rational sustainable development in Vilnius. Among studies focused on water pollution forecasting, we encounter [15,16,17]

The remainder of the paper is organized as follows. Section 2 explains and presents the data, comparatively discusses the evolution of some standard water pollution indicators across G20 and Eastern Europe countries over the last decade and also presents the methodology employed in the econometric investigation. Further, Section 3 discusses the empirical results and model diagnosis, while Section 4 concludes the research.

2. EXPERIMENTAL SECTION

2.1. Data

This study uses annual data on two different water pollution indicators across various countries (both G20 economies and Eastern European economies); first, we employ the total emissions of organic water pollutants (measured in kilograms per day), and second, we use data on water pollution caused by the textile industry (measured in percentages of total BOD emissions) ("Organic water pollutants are measured by biochemical oxygen demand, which refers to the amount of oxygen that bacteria in water will consume in breaking down waste. This is a standard water treatment test for the presence of organic pollutants". (World Bank definition). The period we cover for each time series is from 31 December 1990 to 31 December 2007. Some countries have fewer observations, depending on data availability. The source of the data is the World Bank database.

As is revealed in Figure 1, the country with the highest amount of emissions of organic water pollutants in the world is China, with an impressive value of 9,428,874 kg/day. It is followed by USA (1,850,753 kg/day) and Russia (1,381,683 kg/day) in the group of G20 countries. At the opposite side, *i.e.*, the lowest polluting G20 countries, we find Saudi Arabia (106,621.2 kg/day) and Argentina (155,535.5 kg/day). The Eastern European countries report much lower values for BOD emissions, with the highest values encountered in Poland (359,696 kg per day) and Romania (222,102.50 kg/day); see Figure 2.

Table 1 shows that in Eastern Europe, the percentage of total water pollution that is caused by the textile industry is lowest in the Slovak Republic, Czech Republic and Poland, while in Macedonia, the textile industry is guilty of contributing 44% to this phenomenon. The good news is that compared to the levels achieved in 1990, most Eastern European countries managed to decrease water pollution caused by the textile industry. The only exceptions are Romania (+4.59% as compared to 1990), Bulgaria (+17.24%) and Macedonia (+28.11%), which registered increases in this phenomenon (see Table 1). At the opposite side are Slovenia and the Slovak Republic, both countries managing to decrease water pollution caused by the textile industry by more than 43 percent over the last ten reported years.

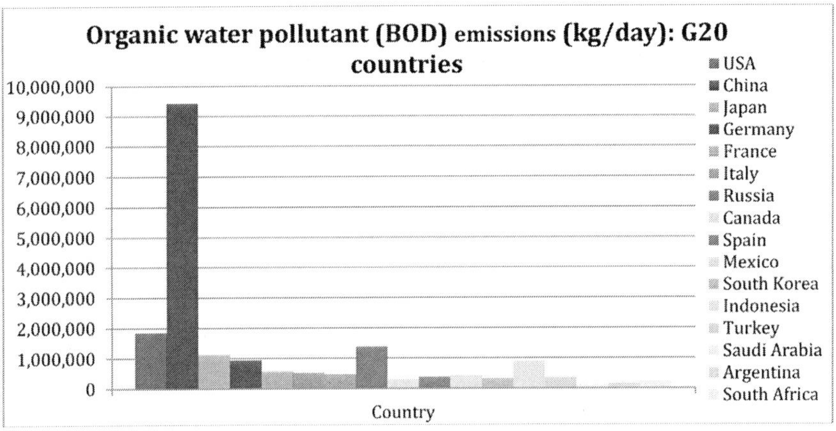

Figure 1. Total emissions of organic water pollutants, G20 countries. BOD, biochemical oxygen demand.

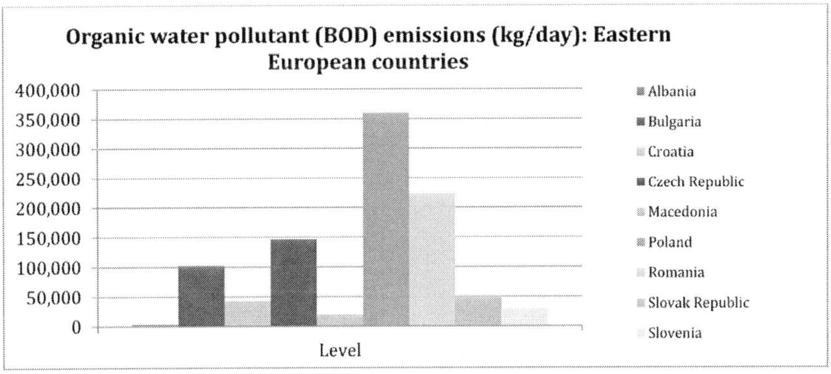

Figure 2. Total emissions of organic water pollutants, Eastern European countries.

Among the more developed G20 economies, Figure 3 reveals that the highest levels for water pollution caused by the textile industry are found in Turkey (32.21%), Indonesia (29.25%) and China (20.29%), while the lowest are in Germany (2.44%), the U.K. (4.26%) and the USA (4.29%). Without exception, all of the G20 countries achieved decreasing rates for their CO_2 emissions in the last reported decade.

Table 1. Water pollution caused by the textile industry, Eastern Europe.

Country	Level	As of	1-Year Change *	~5 Year ago	~10 Year ago
Slovenia	10.79	2007	−0.01%	15.29	19.04
Slovak Republic	5.04	2006	−0.56%	16.42	8.94
Romania	25.04	2007	−0.09%	31.39	23.94
Poland	10.32	2006	−0.06%	12.99	16.18
Macedonia	44.70	2007	0.05%	37.07	34.89
Czech Republic	7.40	2006	0.00%	9.45	12.30
Croatia	14.52	2007	−0.05%	17.71	20.80
Bulgaria	26.79	2007	−0.04%	32.35	22.85

* percentual change relative to last year's value.

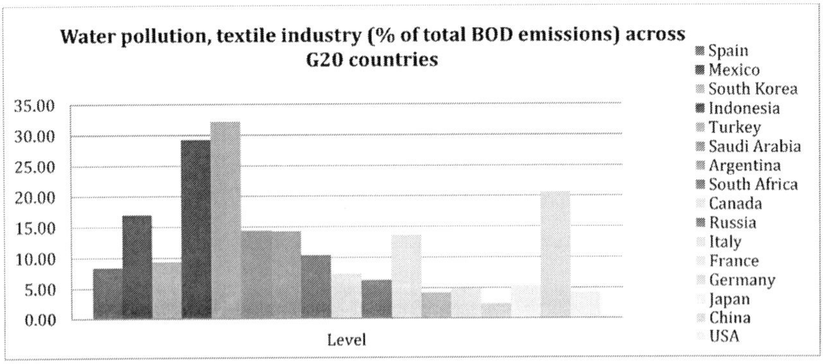

Figure 3. Water pollution caused by the textile industry, G20 countries.

2.2. Methodology

For both Romania and Poland, we make a forecast for the evolution of the two pollution indicators employed in this research, *i.e.*, total emissions of organic water pollutants (measured in kilograms per day) and water pollution caused by the textile industry (as a percentage of BOD) over the next ten years. For this scope, we employ the Holt–Winters model [18,19,20,21]).

The Holt–Winters Model was first proposed in the early 1960s and uses a process known as exponential smoothing. It applies three exponential smoothing formulae to the series, respectively to the mean, trend and each seasonal sub-series, in the following manner:

$$a_t = \alpha(Y_t - s_{t-p}) + (1-\alpha)(a_{t-1} + b_{t-1}) \quad (1)$$

$$b_t = \beta(a_t - a_{t-1}) + (1-\beta)b_{t-1} \quad (2)$$

$$s_t = \gamma(Y_t - a_t) + (1-\gamma)s_{t-p} \quad (3)$$

where:

α, β and γ are the smoothing parameters;

a_t is the smoothed level at time *t*;

b_t is the change in the trend at time *t*;

s_t is the seasonal smoothing at time *t*;

p is the number of seasons per year.

The Holt–Winters forecasts are then calculated using the latest estimates from the appropriate exponential smoothing that has been applied to the series.

Therefore, we have our forecast for the time period T+τ:

$$\hat{y}_{T+\tau} = a_T + \tau\, b_T + s_T \quad (4)$$

where:

aT is the smoothed estimate of the level at time *T*;

bT is the smoothed estimate of the change in the trend value at time *T*;

sT is the smoothed estimate of the appropriate seasonal component at *T*.

3. RESULTS AND DISCUSSION

Figure 4a,b shows that although total emissions of organic water pollutants decreased consistently during the 1990–2007 period in Romania (Figure 4b), the contribution of the textile industry to pollution has increased (Figure 4a). Indeed, water pollution caused by the textile industry in Romania had a constant decreasing trend until 1995, and from 1998, it started to increase sharply, with this upward trend continuing until the year 2004, after which, a reversal occurred, but the last data level is above the initial value.

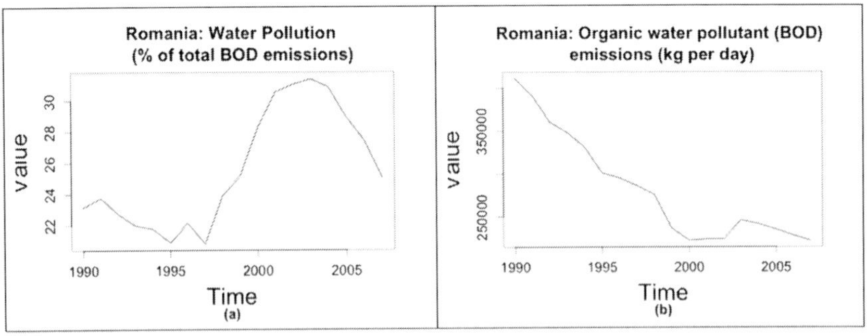

Figure 4. (**a**) Romania: water pollution caused by the textile industry; (**b**) Romania: total emissions of organic water pollutants.

A similar story is true in the case of Poland, with the main difference being that Poland managed to decrease both its total organic water pollutant emissions and the contribution of the textile industry to pollution during the analysis period (Figure 5a,b). Nevertheless, as seen in Section 2, it remains the most polluting country in the Eastern European area. Total emissions of organic water pollutants decreased sharply in Poland until 2002, after which, a reversal occurred (Figure 5b), while the contribution of the country's textile industry to water pollution continued the decreasing trend during the whole analysis period (Figure 5a).

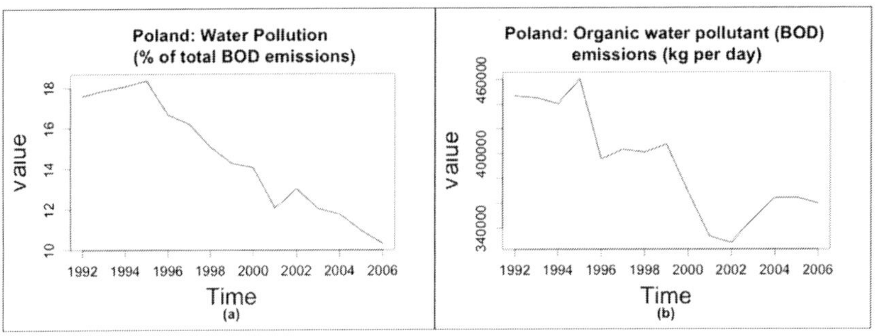

Figure 5. (**a**) Poland: water pollution caused by the textile industry; (**b**) Poland: total emissions of organic water pollutants.

The graphical representation of data in Figure 4 and Figure 5 indicates that this time series appears to be non-seasonal and can be described using an additive model. Thus, for a clearer view of the trend, we will use a smoothing method, by calculating the simple moving average of the time series of order three.

The data smoothed with a simple moving average of order three confirms that organic water pollutant emissions decreased in both countries during the selected period, while the contribution of the textile industry to pollution followed a decreasing trend only in Poland, while Romania registered an increase over the 1998–2004 period.

To make forecasts for both total emissions of organic water pollutants (measured in kilograms per day) and water pollution caused by the textile industry (measured in percentages of total BOD emissions) in Romania and Poland for the following decade, we fit a predictive model using the HoltWinters() function in R, where we set the parameter gamma = FALSE, as our time series appears to be non-seasonal.

Table 2 presents the estimation results for both indictors and both countries.

Table 2. Estimated coefficients of Holt–Winters model.

Coefficients	Total Emissions of Organic Water Pollutants (Kilograms per Day)		Water Pollution Caused by the Textile Industry (% of Total BOD Emissions)	
	POLAND	ROMANIA	POLAND	ROMANIA
a	359,893.8730	222,102.500	10.4199881	25.197353
b	−588.7619	−4,951.716	−0.6754701	−2.287893
Forecasts SSE	848,633.7659	799,484.4617	9.882093	27.83826

Figure 6a–d plots the original time series as a black line, with the forecasted values as a red line. We can observe that the in-sample forecasts agree pretty well with the observed values, although they tend to lag behind the observed values a little bit.

Next, we make forecasts for total organic water pollutant (BOD) emissions and for water pollution caused by the textile industry in Poland and Romania in from 2007–2017. Figure 7a–d reflects the predictions made by the Holt–Winters model as a blue line, with the 80% prediction interval as a dark shaded area and the 95% prediction interval as a lighter shaded area. The decreasing trend is obvious in all cases. According to our estimates, the contribution of the textile industry to water pollution will be 3.66% by the end of the year 2016 in Poland and 2.35% by the end of 2017 in Romania. Both countries are headed towards complete elimination of pollution caused by the textile industry and, hence, towards a sustainable industrial sector, as Greenpeace intended with the release of its reports.

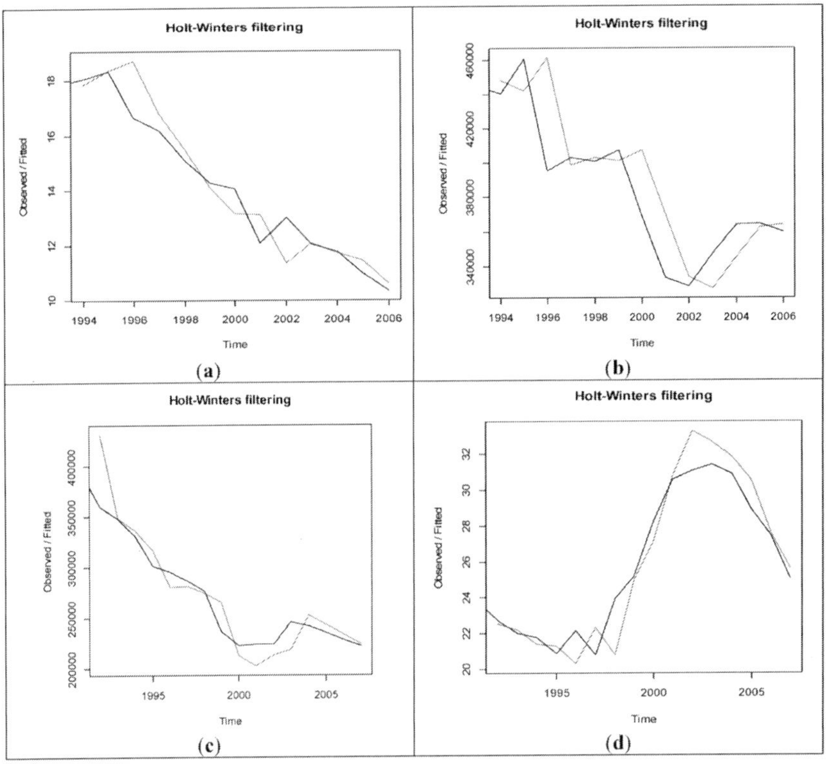

Figure 6. (**a**) Water pollution of the textile industry in Poland; (**b**) Poland: organic water pollutant (BOD) emissions (kg per day); (**c**) Romania: organic water pollutant (BOD) emissions (kg per day); (**d**) water pollution textile industry in Romania.

Finally, we check the accuracy of the forecast by calculating a correlogram of the in-sample forecast errors (Figure 8a–d). We notice from the sample correlograms that the autocorrelations at Lags 1–20 are all within the significance bounds, which proves that our forecast model is correctly specified, as there is no autocorrelation left in the residuals. Further, the Ljung–Box test is carried out in order to test whether there is significant evidence for non-zero correlations at Lags 1–20, and it also confirms the accuracy of the models in all cases (all calculated *p*-values are above 0.6, so we conclude that no statistically significant correlations are left, which is required for a well-specified prediction model). The results therefore attest that both Poland and

Romania are reducing their total organic water pollutant (BOD) emissions and the water pollution caused by the textile industry.

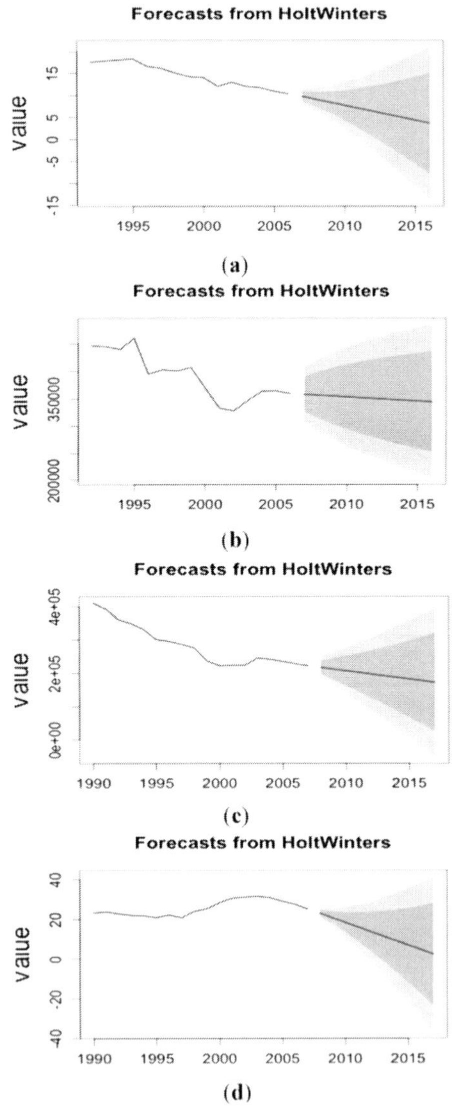

Figure 7. (**a**) Prediction of water pollution caused by the textile industry: Poland; (**b**) prediction of organic water pollutant (BOD) emissions (kg per day): Poland; (**c**) prediction of organic water pollutant (BOD) emissions (kg per day): Romania; (**d**) prediction of water pollution caused by the textile industry: Romania.

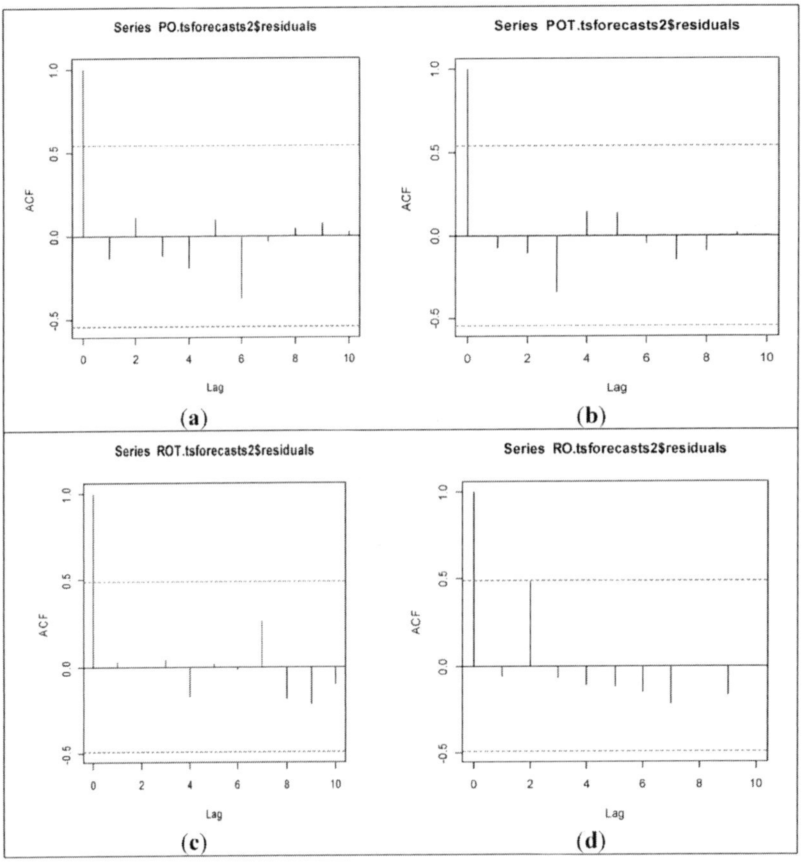

Figure 8. (**a–d**) Correlograms of the in-sample forecast errors for the four time series.

4. CONCLUSIONS

China, the USA and Russia are the biggest emitters of organic water pollutants (BOD) in the world, while Poland and Romania take the first two positions in Eastern Europe. The textile industry is one of the main contributors to water pollution, causing more than 20% of the registered levels of water pollution in countries like Turkey, Indonesia and China (among the G20 group of countries). In Eastern Europe, the percentage of total water pollution that is caused by the textile industry is lowest in the Slovak Republic, Czech Republic and Poland, while in Macedonia, the textile industry is guilty of contributing 44% to this phenomenon. The good news is that compared to the levels achieved in 1990, most Eastern European countries managed to decrease water pollution caused by the textile industry. The only exceptions are Romania (+4.59% as compared to 1990), Bulgaria (+17.24%) and Macedonia (+28.11%), which registered increases in this phenomenon.

Given the controversy created by the textile industry's contribution to pollution at a global level, this paper forecasts the trend of the total emissions of organic water pollutants, as well as of the textile industry's contribution to pollution over the next decade for the two Eastern European countries included in the analysis, *i.e.*, the top polluters Poland and Romania. Predictions made with the Holt–Winters forecasting model reflect a continuous decrease of both pollution indicators in both countries, with a point estimate of the contribution of the textile industry to water pollution of 3.66% by the end of the year 2016 in Poland and of 2.35% by the end of 2017 in Romania. All models are correctly specified as attested by standard econometric tests, which allows us to confirm that both countries are headed towards almost complete elimination of pollution caused by the textile industry in the near future.

Corroborated by the fact that, in general, any action plan to foster sustainable development will need to include pollution reduction measures, we conclude that the Eastern European countries analyzed in this research are on their way towards a more sustainable industrial sector.

ACKNOWLEDGMENTS

This research was supported by CNCS-UEFISCDI (National University Research Council- Executive Agency for Higher Education, Research, Development and Innovation), Project Number IDEI 303, Code PN-II-ID-PCE-2011-3-0593.

AUTHOR CONTRIBUTIONS

The three authors contributed equally to Section 1, Section 3 and Section 4, while Cristiana Tudor contributed to Section 2. Cristiana Tudor prepared the manuscript and all authors were involved in discussing the study. All authors have read and approved the manuscript.

CONFLICTS OF INTEREST

The authors declare no conflict of interest.

REFERENCES

1. Dirty Laundry: Unravelling the corporate connections to toxic water pollution in China. Available online: http://www.greenpeace.org/international/Global/international/publications/toxics/Water%202011/dirty-laundry-12pages.pdf(accessed on 5 September 2014).

2. Clothing to dye for: The textile sector must confront water risks. Available online: http://www.theguardian.com/sustainable-business/dyeing-textile-sector-water-risks-adidas (accessed on 5 September 2014).

3. Parvathi, C.; Maruthavanan, T.; Prakash, C. Environmental impacts of textile industries. *Indian Text. J.* **2009**, *CXVII*, 22–26.

4. Dirty Laundry 2: Hung Out to Dry. Unravelling the toxic trail from pipes to products. Available online: http://www.greenpeace.org/international/Global/international/publications/toxics/Water%202011/dirty-laundry-report-2.pdf(accessed on 5 September 2014).

5. Tudor, C. Modeling the Evolution of the Romanian Textile Industry: Empirical Evidence of Lead-lag and Causal Relationships. *Fibres Text. East. Eur.* **2012**, *20*, 8–12.

6. Domańska, D.; Wojtylak, M. Application of fuzzy time series models for forecasting pollution concentrations. *Expert Syst. Appl.* **2012**, *39*, 7673–7679.

7. Osowski, S.; Garanty, K. Forecasting of the daily meteorological pollution using wavelets and support vector machine.*Eng. Appl. Artif. Intell.* **2007**, *20*, 145–155.

8. Niska, H.; Hiltunen, T.; Karppinen, A.; Ruuskanen, J.; Kolehmainen, M. Evolving the neural network model for forecasting air pollution time series. *Eng. Appl. Artif. Intell.* **2004**, *17*, 159–167.

9. Ziomas, I.C.; Melas, D.; Zerefos, C.S.; Bais, A.F.; Paliatsos, A. Forecasting peak pollutant levels from meteorological variables. *Atmos. Environ.* **1995**, *29*, 3703–3711.

10. Sahu, S.; Mardia, K.V. A Bayesian kriged Kalman model for short-term forecasting of air pollution levels. *J. R. Stat. Soc. Ser. C (Appl. Stat.)* **2005**, *54*, 223–244.

11. Kurt, A.; Gulbagci, B.; Karaca, F.; Alagha, O. An online air pollution forecasting system using neural networks.*Environ. Int.* **2008**, *34*, 592–598.

12. Ibarra-Berastegi, G.; Elias, A.; Barona, A.; Saenz, J.; Ezcurra, A.; Argandoña, J. From diagnosis to prognosis for forecasting air pollution using neural networks: Air pollution monitoring in Bilbao. *Environ. Model. Softw.* **2008**, *23*, 622–637.

13. McCollister, G.M.; Wilson, K.R. Linear stochastic models for forecasting daily maxima and hourly concentrations of air pollutants. *Atmos. Environ.* **1975**, *9*, 417–423.

14. Zavadskas, E.K.; Kaklauskas, A.; Kaklauskiene, J. Modelling and forecasting of a rational and sustainable development of Vilnius: Emphasis on pollution. *Environ. Pollut.* **2007**, *30*, 485–500.

15. Singh, K.P.; Basant, A.; Malik, A.; Jain, G. Artificial neural network modeling of the river water quality—A case study.*Ecol. Model.* **2009**, *220*, 888–895.

16. Rauch, W.; Harremoës, P. The importance of the treatment plant performance during rain to acute water pollution.*Water Sci. Technol.* **1996**, *34*, 1–8.

17. Zhu, Z.; Deng, Q.; Zhou, H.; Ouyang, T.; Kuang, Y.; Huang, N.; Qiao, Y. Water pollution and degradation in Pearl river Delta, South China. *AMBIO A J. Hum. Environ.* **2002**, *31*, 226–230.

18. Chatfield, C. The Holt-Winters Forecasting Procedure. *Appl. Stat.* **1978**, *27*, 264–279.

19. Gelper, S.; Fried, R.; Croux, C. Robust forecasting with exponential and Holt-Winters smoothing. *J. Forecast.* **2010**, *29*, 285–300.

20. Holt, C.C. Forecasting Trends and Seasonals by Exponentially Weighted Averages. *Int. J. Forecast.* **2004**, *20*, 5–10.

21. Lynwood, A.; Johnson, D.; Montgomery, C.; Gardiner, J.C. *Forecasting and Time Series Analysis*, 2nd ed.; McGraw-Hill Inc: New York, NY, USA, 1990.

10

THERMOSET COMPOSITES REINFORCED WITH RECYCLED COTTON TEXTILE RESIDUES

Welton Fernando Zonatti* , Bárbara Maria Gama Guimarães, Wânia Duleba and Júlia Baruque Ramos

University of São Paulo, School of Arts, Sciences and Humanities, Av. Arlindo Béttio, 1000, 03828-000 São Paulo, Brazil

ABSTRACT

The recycling of textiles is an issue that requires immediate attention in order to address the management of textiles derived from household waste, as well as scraps generated throughout manufacturing textile processes. The present study aimed to produce composites employing three different types of thermoset resins (epoxy, polyester orthophthalic and polyurethane) reinforced with recycled cotton textile residues on which tensile tests were carried out. The incorporation of 30% (v/v) of textile reinforcement in composites made from the specific epoxy resin employed in the present study showed an increase of approximately 2-fold relative to pure resin tenacity and Young's modulus (respectively from 0.09 to 0.16 MPa and from 0.5 to 1.2 MPa). For composites made of orthophthalic polyester resin, an increase was observed of 2-fold relative to pure resin tenacity and 3-fold relative to Young's modulus (respectively from 6.5 to 13.8 MPa and from 193 to 772 MPa). No significant alterations were observed for these parameters in polyurethane resin composites (16 MPa for tenacity and near 700 MPa for Young's modulus) probably because the modulus value supplied by the fraction of cotton fiber is not as significant as that one from the pure polyurethane resin. Two propositions of fashion accessories were made by employing these composite materials. Also, the artworks of renowned Brazilian designers, namely, Campana Brothers and R. Sobral, were analyzed to develop the artwork of such accessories for the present "documentation process". Thus in the present study aesthetic and technical aspects are analyzed together covering all production phases of material, processes, creation and design of final product.

KEYWORDS

Composite Thermoset, resin, Recycled cotton residues, Sustainable design Fashion

1. BACKGROUND

In Brazil large amounts of textile clothing (often in good conditions) are being discarded along with scraps and textile virgin residues originated by apparel confection industries. Through enticing advertisements in virtual media such as newspapers, magazines, Internet and especially television, fashion can generate new aesthetic needs and desires.

According to a survey by MDIC - Brazilian Ministry of Development, Industry and External Trade, and ABIT - Brazilian Association of Textile and Apparel Industry (2011), carried out in 2010, television was pointed as the major medium of fashion dissemination by 72% of respondents. Furthermore, 47.5% of them have already bought fashion products as a consequence of watching advertisements. Thus, consumption is stimulated, offering the false satisfaction of desires by the continued purchase of goods. Added to this strategy of the fashion industry, there is also planned obsolescence (Queiroz 2003; Broega et al. 2011, in which the life of a textile is related to a predetermined period of time, being discarded after this period - often in good conditions - indiscriminately by consumers.

On the other hand, there are also large amounts of scraps and textile residues originated by apparel confection industries. In Brazil, there are many regions with intense confection activity, mainly in the states of São Paulo, Santa Catarina and some parts of the Northeast region of the country. According SINDITEXTIL - Union of Spinning and Weaving Industries of the State of São Paulo – Brazil (2012), Brazil produces 175,000 ton of residues from the textile industry per year. In São Paulo city (São Paulo state, Brazil), approximately 10 ton per day (in the neighborhood of Brás) and 16 ton per day (in the neighborhood of Bom Retiro) of textile residues are produced and usually this waste ends up in landfills (Vianna 2011).

Considering one example in particular, in order to delineate the problem's dimension on textile residues generation, data from the neighborhood of Bom Retiro were focused. This neighborhood is one of the largest commercial fashion centers in Brazil. The confection producers in this region lay down bags of fabric scraps on the streets; afterwards, garbage pickers tear the bags to collect material, leaving a lot of thrown scattered trash (Leite 2011). The region has 1,200 installed confection industries and the collection of textile wastes is poorly structured (CNI – Brazilian National Confederation of Industries, and ABIT – Brazilian Association of Textile and Apparel Industry 2012).

Even with a large number of cooperatives working in the sector, the major part of this waste is not reused or recycled due to issues with logistics, transportation and lack of infrastructure. Thus, the neighborhood of Bom Retiro is the greatest generator of textile waste of São Paulo city. A large amount of this waste are fabrics made of cotton, the most consumed raw textile material in Brazil (nearly 70% of the total amount) (Costa and Rocha, 2009). The recycling of cotton could bring great benefits to the environment regarding sustainability in the textile and fashion industry. The production and consumption of a simple cotton shirt with 250 g (0.55 lb) implies in the consumption of 1.7 kg (3.75 lb) of fossil fuels, 450 g (1.0 lb) of waste in landfills and the issuance of 4 kg (8.82 lb) of CO_2 to atmosphere. Therefore, the reuse and recycling of fibers provide environmental and economic benefits, for instance (Alwood et al. 2006):

- Reduction of landfill space. Textiles in landfills present, among other matters, problems such as the remains of synthetic fibers, which take a long time to disintegrate. Also, woolen garments when decomposing produce methane, which contributes to global warming;
- Reduction of pressure on natural resources;
- Satisfactory results in pollution control and energy economy;
- Reduction of spending on raw materials.

One possibility for recycled fibers could be to employ them as reinforcement in composite materials, which in turn could be used in the development of design and fashion products. Composite materials have at least two components or two stages in which physical and chemical properties are clearly different between them. Separately, these constituents maintain their characteristics, but when mixed they form a composite of which properties are impossible to obtain with just one of them. The properties of the composites can be considered as a combination between the properties of the matrix and fibers in addition to the properties of interfaces between the fibers and matrix (Carvalho 2005; Saville, 2007).

In order to find possibilities of recycling, the present study aimed to produce composites employing three different types of thermoset resins (epoxy, polyester orthophthalic and polyurethane) reinforced with recycled textile cotton fibers. Their resistance characteristics were evaluated by means of tensile tests. This study also attempts to conduct a proposal of fashion product application, suggesting a profitable destination different from ordinary discard in landfills.

2. METHODS

Residues constituted by scraps and cotton yarns were originated by tearing a post-consumption jeans (denim) trouser (M. Officer, model Five Pockets, Brazil), which fabric is from Tavex Corporation company (Americana city, state of São Paulo, Brazil). This fabric was composed of cotton (100%), heavy weight 12 oz yd^{-2} (about 410 g m^{-2}), twill structure (3x1), 67 yarns of warp x 41yarns of weave per square inch, warp dyed with indigo colorant (6.5 Ne – where Ne means the number of hanks per pound – explaining that 1 hank corresponds to 840 yards or 768 meters per pound or 0.454 kg – corresponding to 90 Tex (90 g per 1,000 m of yarn), weave (or filling) undyed (9.0 Ne or 65.6 Tex), produced from carded yarn by open-end spinning system (Figure 1 and Figure 2a).

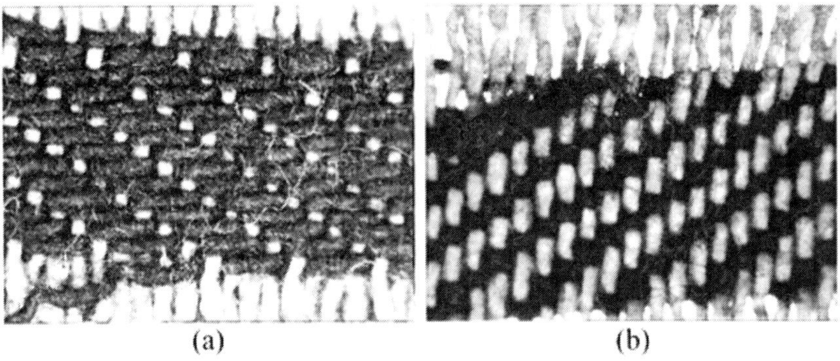

Figure 1. Texture of jeans (denim) fabric. Texture of jeans (denim) fabric made of cotton (100%), heavy weight 12 oz yd^{-2} (about 410 g m^{-2}), twill structure (3 × 1). **(a)** Texture of front of the fabric with a focus on dyed warp (×32 magnification) and **(b)** texture of the back of fabric with a focus on undyed weave (or filling) (×32 magnification).

The trouser was shredded with the aid of scissors into pieces of approximately 5x5 cm and teared in a blender (model Due Sapore, Britannia, Curitiba/Paraná, Brazil) in order to simulate the process on a small scale of rotary blades of an industrial textile tearing plant. The obtained result is supposed to be similar to that obtained by means of specific tearing machines for processing textile waste. The length of this reinforcement made from recycled cotton fabric varied from 2 to 3 cm.

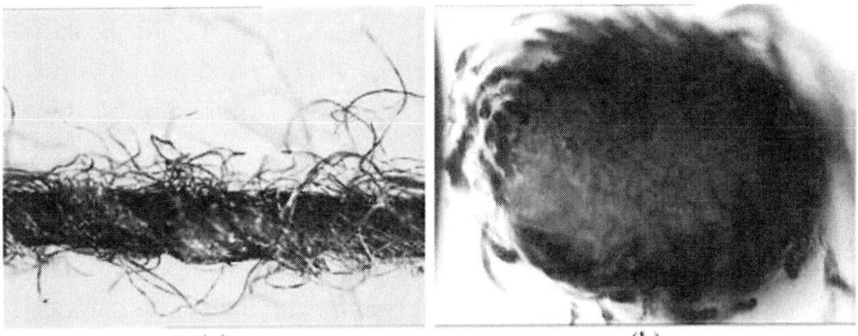

(a) (b)

Figure 2. Longitudinal structure and cross microscopy of cotton open-end warp yarn. (a) Longitudinal structure of cotton open-end warp yarn (magnification ×51). **(b)** Cross microscopy of cotton open-end warp yarn superficially indigo dyed. Inside the presence of undyed cotton fibers is noted (magnification ×320).

In order to determine the moisture recovery (or regain) of textile reinforcement, according to the standard ABNT, ISO 139:2005 – Brazilian Association of Technical Standards and International Organization for Standardization 2005 and ISO/TR 6741–4–1987 - International Standards Organization / Technical Report 1987, the obtained textile residues were conditioned for a minimum period of 48 hours at 20°C and relative humidity of 65% (Climatest model M250-RH, Mesdan, Raffa di Puegnago/Brescia, Italy). Subsequently, the amount of moisture was determined by weighing a conditioned sample in an analytical balance (model ED124S, Sartorius, Göttingen, Germany) and then drying was carried out in an oven with forced air circulation (model FD 115, Binder, Tuttlingen, Germany) at 70°C until achieving constant weight (overnight) and then the sample was weighted again. The moisture recovery (or regain) is defined as the weight of water (original weight minus dry weight) calculated as a percentage of the dry weight.

For matrix composition of composites the following were employed: i) crystalline epoxy flexible resin BVR 222 (Bonivitta, São Paulo/SP, Brazil) and hardener BVE 0101 (Bonivitta, Brazil); ii) crystalline resin Arazyn 1.0 (polyester orthophthalic) (Artsol, Sumaré/SP, Brazil) and hardener T-Catalyst (Moldflex, Brazil); iii) crystalline polyurethane resin of low viscosity Poly-Optic 14 (Moldflex, Brazil) and hardener T-Catalyst (Moldflex, Brazil).

The reinforcement composed by recycled fabric cotton residues was placed at the bottom of a container mold (H = 9.7 cm; L = 28.5 cm; W = 17.6 cm) and covered by resin. The specimens (plate composites) consisted of approximately 30% of reinforcement (apparent volume of 80 mL) and the remaining volume of resin (epoxy, polyester and polyurethane), which was approximately 200 mL. Regarding weight, approximately 8.5 g of residues were employed. The polymeric matrices had density of 1.1 g cm^{-3}, near 1:25 (weight/weight ratio). The final thickness of the samples was approximately 5 mm. The employed equipment and conditions for composite development were respectively: an open container, temperature near 25°C (room temperature), normal atmospheric pressure and 48 hours of resting in order to ensure the proper curing.

Once hardened, the composites were sliced with the aid of a band saw. Ten samples of each of the three different kinds of resins reinforced with fiber and ten samples of each of the three different types of resins without any reinforcement were tested mechanically. The obtained results were compared. Tensile tests were performed according to ABNT, NBR 13041:2004 - Brazilian Association of Technical Standards - Brazilian Standard 2004 (determination of tensile strength and elongation) employing tester machine Instron (model 5569, Norwood, MA, USA).

In order to observe the fabric, microscopies in stereo-microscope (Leica, MS5 model, Germany) coupled to video camera to capture digital images (model VPC 122/CH, 1/2 in. CCD, Vista Protos IV, Wokingham/ Berkshire, UK) were performed. For that, the longitudinal yarn structure was affixed onto a cardboard frame and the dispersion and anchoring of the cotton reinforcement in various polymer matrices (epoxy, polyester and polyurethane) were tested, as well as the behavior of the anchoring of the reinforcement in the matrices near the sites of rupture after tester machine tests. The magnifications were 20, 32, 51, 80 and 128 times. The images were captured and processed by the system code 250 Video Analyser 2000 (Mesdan, Italy).

In order to obtain a sectional cross view of the cotton yarn, such was encapsulated in microscopy resin (Technovit 7100, Heraeus, Germany). The cross sections were performed in rotational semi-automated microtome (Leica, model RM 2245, Germany) equipped with cut knife type "C". The slices were placed on microscope glass slides, flattened with a drop of petrolatum mineral oil (Nujol, Mantecorp, Rio de Janeiro/RJ, Brazil) and covered with glass cover slips. They were analyzed in a biological microscope (Leica, Model BME, Germany) coupled to a video camera to capture digital images (Sony Color Video Camera, model P-55C-DC93, Sony, Shanghai, China). The magnifications were of 128 and 320 times. The images were captured and processed by the system code 250 Video Analyser 2000 (Mesdan, Italy).

Silicone resin RTV-2 (Moldflex, Brazil) and catalyst T (Moldflex, Brazil) in a proportion of 4% relative to the total volume were employed in order to make the molds for the accessories development (bracelet and ring). Preexisting jewelry gems and bracelets were employed as pre-molds. They were affixed to the bottom of the cardboard container with highly adhesive glue (Loctite - Super Bonder, Henkel Brazil, São Paulo/SP, Brazil. The silicone resin was then mixed with the catalyst and spilled into the cardboard container that was previously sealed with insulating tape to avoid any liquid silicone resin leak.

After five hours, at room temperature and normal atmospheric pressure, the resin was cured, the cardboard container (bottom part) was teared and together the jewelry pre-molds were removed from the negative silicone molds formed. The upper part of these negative molds received a silicone layer of at least one centimeter in order to support and join the necessary negative molds to produce gems and bracelets positive shapes. The bracelet was made of epoxy resin reinforced with textile fibers. The ring gem had the same material as the bracelet, however a small portion of cut golden metalized ribbon was added, in order to provide a glitter effect.

3. RESULTS AND DISCUSSION

3.1 Cotton recycled reinforcement

The analysis of longitudinal yarn structure aimed to confirm the production process of the yarn employed in this study, due to the fact that denim can be produced through a ring spinning system or rotor spinning (open-end) (Figure 2a). This last one is the most common system for producing yarn for denim weaving (Adanur, 2001; Kaswell 1963).

The rotor spinning, also known as open-end spinning, is used for wiring very short length fibers and it is a generic term for the production of yarn from staple fibers. The term can be employed for any method in which the tape or roving is opened or separated into its individual fibers or tufts, and then reconstituted in the spinning device to form the yarn.

Originally, denim was only made of cotton. Afterwards, this fabric started to be made also with a blend of cotton and polyester, despite the fact that the dyeing of polyester or blended warp is more difficult than 100% cotton warp. However, the employment of polyester fiber is often as manner to reduce the final cost of the product (Adanur, 2001).

By means of cross section microscopies (Figure 2b), which were compared with reported cotton cross microscopy (Mesdan-Lab 2005), it was possible to identify that the yarn employed in this study is of 100% cellulosic cotton natural originate, since it is possible to observe in the microscopy the typical transversal shape of cotton fiber with the lumen (small orifice, characteristic of natural fibers) inside it. Similarly, the same was observed for the various fibers composing the cotton yarn (which can be also a cable composed of more than one yarn), the main raw material employed for the production of denim fabric. The indigo dye does not have high affinity for the cotton. As a result, the dying is only superficial on warp yarns, forming a blue ring and leaving a white core, leading to a low fabric color fastness under friction (Marsh et al., 2003).

According to Adanur (2001), the open-end is a process notably employed to produce denim yarn. The open-end yarns are cheap and rustic for not passing through machines that add a sophisticated finishing, such as textile laminating and combing machines, differing, for example, from the ring spinning process where a set of parallelized fibers is well twisted so there is high cohesion and strength. Additionally, in open-end spinning, the individual fibers are not uniformly arranged in the produced yarn as in the spinning ring produced yarn. In an open-end spinning yarn most of the fibers are parallel, but with some intersections and some irregular fibers surrounding the main cable (Fabriclink – Fabric University 2012).

The longitudinal view of the cotton yarn employed in the present study (Figure 2a) shows that it was produced by an open-end process, since it is irregular and with a large number of fibrils, various thinner loose fibers throughout the yarn, presenting also some intersections between themselves and containing some irregular fibers surrounding the main cable.

According to Saville (2007) and Kaswell (1963), the percentage of moisture recovery (or regain) of cotton fiber is around 8.5%. However, after performing this test with 3 replicates, it was found that for the recycled textile reinforcement employed in this study, the values were 9.1%, 9.4% and 10.0%, resulting in $9.5 \pm 0.5\%$ (average and standard deviation). The accuracy of the analysis is demonstrated by the variation coefficient of only 5%.

Thus, probably the denim fabric employed in the present study was subjected during its production to a chemical treatment to confer it a softer touch, which at the same time increased its hydrophilicity and, possibly, its affinity to the indigo dye. As a hypothesis, the denim fabric may have been subjected to mercerization in the secondary stage of textile processing. The process of mercerizing consists in applying cold concentrated caustic soda (sodium hydroxide - NaOH) on the tensioned fabric. The caustic soda reacts with the cellulose fibers of the cotton causing a swelling, leaving them with a more rounded profile and decreasing the amorphous regions of the cellulose. The regain of mercerized cotton is approximately from 10 to 12% (Saville, 2007; Kolpak et al. 1978).

Microscopic transversal examinations of the fibers with magnification of 500x or more allow to determine if the cotton fibers are mercerized or not (Saville, 2007; Kaswell 1963). In the case of the fabric employed in this study, in the eventual case of mercerization process, it possibly would not be applied primarily on the fibers, but directly on the tensioned fabric in the secondary stage of textile processing. The carried out optical transversal microscopies, such as shown in Figure 2b, presented transversal profiles compatible to that expected to non mercerized cotton. However, the observation aforementioned cannot be conclusive, since for that the ideal would be to compare the obtained microscopy images with pre-existing patterns (before and after mercerization process) of the same cotton fibers employed for producing the denim fabric employed in this study. On the other hand, the mercerization employ for denim cotton fabric manufacturing is not usual, because it is a prolonged process that requires specific machinery (Santanaense 2014).

As previously mentioned, probably the denim employed in the present study has not been subjected to mercerization process. Nevertheless an example of premium mercerized denim was introduced in the market in 2012 by Santana Textiles manufacturer (Brazil), owner of "Loco Serious Denim" brand (Santana Textiles 2014). This mercerized denim line was called "Absynto" (100% cotton, twill right structure 3x1, width: 1.62 m, weight: 10.80 oz yd^{-2}, color: intense blue, finishing: flat) (Bergamasco 2012).

Another more plausible possibility is that, regardless of being or not mercerized, the denim fabric has received in tertiary or late stage of processing, a cellulase enzyme treatment, which provided a controlled hydrolysis or change of the structure of cellulose molecules - the main cotton constituent component (Basra and Malik 1984). Cellulases are able to make cotton fabrics smoother and softer, degrading the surface fibers (loose fibers and microfibrils), conferring also a stone washed appearance (or the aspect of "wear after washings") in jeans (Ferrante 2011; Wadaa et al. 2010).

At last hypothesis, detergents with alkaline pH formulations are normally used to launder clothes (Nascimento and Martins 2006). In this way, taking in account that in this study a post-consumption trouser made of denim was employed, the cotton fibers of this fabric could be affected by exposure to this alkalinity after repeated washings in alkaline pH condition.

Nonetheless, according to the obtained regain value observed it is possible to infer that the fabric, which originates the textile reinforcement employed in this study, has received a treatment during its manufacture, which increased the cotton hydrophilicity, meaning also that the crystallinity of cellulose structures were diminished and the amorphous ones were increased (Reddy and Yang 2005).

3.2 Composites production and analysis

Firstly, it is worth pointing that despite the low values of tenacity and modulus presented in Table 1 for epoxy resin, according to information provided by the supplier (Bonivitta, Brazil), this resin is 100% epoxy resin, since epoxy is a designation of a large family of resins. There are different types of epoxy for different kind of applications (with correspondent different degrees of hardness, toughness, cost, etc.). An epoxy or polyepoxide is a thermosetting plastic that hardens when mixed with a catalyst or "hardening" agent. The most common epoxy resins are products of a reaction between epichlorohydrin and bisphenol-A.

Currently epoxy resins are used for a multitude of applications, such as the inner lining packaging of beer, soda and citrus, among others, printed circuit boards, computer motherboards, encapsulation of electronic components, flash drives, industrial floors, surfboards, anticorrosive coatings, powder painting, among other applications that also use epoxy as a base. Thermosetting resins (or thermosets) play an important role in industry due to their high flexibility for tailoring desired ultimate properties, leading to their high modulus, strength, durability, and thermal and chemical resistances as provided by high cross-linking density. There are now several companies worldwide that specialize in formulating products with their own characteristics attending the need of each user (Raqueza et al. 2010).

Ortophtalic unsaturated polyester resins have in their composition phthalic acid, also known as orthophthalic acid, or anhydride, which dissolves itself in alcohol and some other solvents. This type of resin is the most common, because its use is widespread and low cost. The mechanical properties of ortophtalic unsaturated polyester resins are low. These resins can be used in the production of buttons and toilet seats, and can also be used in the production of electronics, such as the manufacture of reactor lamps. Other suggestions for usage are: casting handicraft and decorative objects as panels, stained glass, table tops, jewelry, statues, trophies, casting parts for toilets and toilet seats, soap dishes, trash cans, sinks and towel racks. Along with glass fiber, it can be used in composing construction tiles (manual process), domes and shutters (Poliresinas 2012; Amico 2010).

Polyurethane resin can be defined as any resin resulting from the reaction of diisocyanates (such as toluene diisocyanate) with an amine, phenol or a carboxylic or hydroxylic compound to produce a polymer with free isocyanate groups that can be employed for prototyping applications, protective coatings, potting, casting resins, adhesives, rubbers, foams, and in paints, varnishes and adhesives (Gil 2009; Zia et al. 2007).

In tests, epoxy resin, polyester and polyurethane were used in pure samples and samples reinforced with textile waste (Figure 3). According to the results shown in Table 1 in all cases it was observed that the resistance of all materials increased with the addition of textile reinforcement: in the case of composites made from epoxy resin, an increase of approximately 2-fold relative to pure without reinforcement resin tenacity and modulus (respectively from 0.09 to 0.16 MPa and from 0.5 to 1.2 MPa). For composites made of orthophthalic polyester resin, those ones with fiber reinforcement showed approximately 2-fold relative to pure resin for tenacity and 3-fold more for module values (respectively from 6.5 to 13.8 MPa and from 193 to 772 MPa).

Table 1. Tensile results obtained in tester machine assays

	Maximum load (N)	Tenacity (MPa)	Extension break (mm)	Young's modulus (MPa)
Pure epoxy resin	9 ± 1 (11%)	0.09 ± 0.01 (11%)	35 ± 4 (11%)	0.5 ± 0.1 (23%)
Epoxy resin + cotton fibers	17 ± 1 (7%)	0.16 ± 0.01 (7%)	21 ± 4 (17%)	1.2 ± 0.2 (13%)
Pure polyester resin	654 ± 58 (9%)	6.5 ± 0.6 (9%)	5.6 ± 0.4 (7%)	193 ± 19 (10%)
Polyester resin + cotton fibers	1,379 ± 72 (5%)	13.8 ± 0.7 (5%)	2.7 ± 0.3 (10%)	772 ± 120 (16%)
Pure polyurethane resin	1,607 ± 408 (25%)	16 ± 4 (25%)	3.6 ± 0.5 (15%)	679 ± 152 (22%)
Polyurethane resin + cotton fibers	1,618 ± 427 (26%)	16 ± 4 (26%)	4.0 ± 0.8 (20%)	702 ± 225 (32%)

Tensile results obtained in tester machine assays for epoxy, polyester, and polyurethane resin not reinforced with cotton residues and, respectively, composites reinforced with cotton residues. The values are expressed as average, standard deviation and, in parentheses, the coefficient of variation.

Figure 3. Composite made of polyurethane resin specimen. Strip cut for tensile testing in tester machine with 2 cm wide, 10 cm of effective length plus 3 cm test at each end.

Only for polyurethane resin composites the incorporation of textile recycled reinforcement did not imply in significant alterations in tenacity and module values (16 MPa for tenacity and near 700 MPa for Young's modulus). A plausible explanation for this is to consider that the modulus value supplied by the fraction of cotton fiber (reinforcement) not as significant as that one from the pure polyurethane resin (matrix). Taking in account that the cotton fiber modulus ranges from 8.2 to 9.6 GPa (adapted values from Reddy and Yang2005; Kaswell 1963) and, as a simple approach that the final modulus is proportional to the moduli of constituents (matrix and reinforcement), the resulting modulus of the polyurethane composite would be: $0.96\times(0.679) + 0.04\times (9.0) = 1.011 \sim 1$ GPa, which is within the error range of the made measurements (Table1).

Thus, the cotton fiber in the proportion that has been employed in relation to the total polyurethane resin, provided only a small increase in the module value, but not as significant as that observed in the composites made with epoxy or polyester resins. It should also be noted that in the preparation of the composite, shattered yarns instead of totally separate cotton fibers were employed. So it is possible to cogitate that the correspondent modulus value for these shattered yarns was inferior to that of isolated cotton fibers.

It was observed that the values of the modulus for the epoxy resin specimens with and without fiber reinforcement are compatible to those for foams (from 0.001 to 0.01 GPa) and rubbers (from 0.01 to 0.1 GPa) (UTL - Technical University of Lisbon 2012); while the values of the modules for the samples of orthophthalic polyester resin and polyurethane, with and without reinforcement of fibers, are closer to the ones for polyethylene (from 2.5 to 4.0 GPa) and nylon as plastic material (from 1.5 to 4.0 GPa) (Callister2007).

Nevertheless, it has been noted that the experimental values of moduli shown in Table 1 are smaller than the values of the polyethylene and nylon polymers (engineering plastics). This finding is consistent with the fact that the assayed resins are currently employed in handicrafts and customized production of articles, which normally present smaller resistance values than the routinely applications of engineering plastics. However for fashion or design applications generally high values of resistance parameters are not necessary as in the case of engineering applications, being more relevant the aesthetic appearance.

Withal, the main limitation about incorporation of natural fibers in polymeric materials is relative to the loss of mechanical resistance, many times due to the weak interaction between the polymer and the fiber (Furlan et al. 2012), since the lack of cohesion influences directly the intermaterial tensions. In some cases, composites could have the tendency of forming voids and cracks in the interior and on the surface of its components (Almeida and Rezende 2001).

In the present study, the good adhesion (interface) of the two materials – textile reinforcement and resins - was evidenced by the increase of resistance and Young's modulus values (Table 1). It was also possible to verify the good adhesion of all produced composites by the microscopy magnifications (32x) (Figure 4), since torn fibers within the polymer matrix were not found in the places of rupture of the composites after testing in the tester machine, nor cracks or fractures in the composite material as a whole, indicating the proper anchor between reinforcement and matrix. Figure 4 also shows the dispersion of microfibrils inside all three produced composites, being probable that they also aided to increase the good adhesion between matrices and reinforcement.

Taking in account these findings, a first consideration would be that the intersections and some irregular fibers surrounding the main cable of open-end yarn could have promoted the presence of fibrils and microfibrils in the final composite, which probably could increase the good adhesion between matrices and reinforcement. On the other hand, taking into account the previous cited shredding process for textile reinforcement production, it is not possible to ensure the integrity of the yarns originally present in the fabric inside the final composite. Hence, a most plausible hypothesis is to consider that the majority of fibrils and microfibrils were originated from shredding process for production of the textile reinforcement rather than as a consequence of the structure of the open-end yarn.

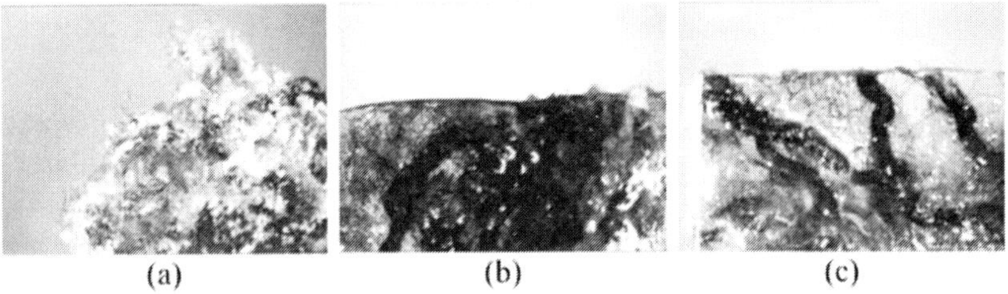

Figure 4. Magnifications of 32 times in the site of the composite failure after tester machine assays. These confirm good cohesion between the polymer matrices and textile waste reinforcement. **(a)** Epoxy resin composite. **(b)** Orthophthalic polyester resin composite. **(c)** Polyurethane resin composite.

Concerning the blue colored warp, visible to the human eye or by microscopy magnifications (32x) (Figure 4), indigo dye diffusion from the yarn surface to the matrices was not observed. It is worthy of mention that the present authors did not find in literature reports about these questions and neither about the employment of composite reinforcement from raw textile material originally composed by colored or not open-end, spinning ring or other kind of yarn.

Aiming to increase the interfacial compatibility between lignocellulosic fibers (hydrophilic) and polymer matrices (usually hydrophobic) in composites, various chemical treatments are usually performed mainly targeting the following modifications on the fiber surface: removal of non-cellulosic components (mainly lignin, hemicellulose and residues), increasing of hydrophobicity and increasing of surface roughness (Carneiro 2013). In addition, it is reported that hydrophilic cellulosic fibers do not adhere well in hydrophobic polymer matrices (Rosario et al. 2011).

In the present study, cotton textile residues were employed as reinforcement. Cotton is a material almost totally constituted by cellulose (Saville, 2007; Kaswell 1963). This molecule has a strong tendency to form inter and intramolecular hydrogen bonds, mainly due to hydroxyl groups. The hydrogen bonds and Van der Waals forces between the various planes are responsible for adding fibrils and microfibrils, allowing it to form an ordered and compact structure which constitutes the crystalline structure of cellulose. However, the hydrogen bonds not only with the cellulosic hydroxyl chain, but also with water. Thus cellulose has great affinity with water, despite being insoluble in it (Sampaio 2013).

The regain (or moisture content) determination is an indirect way to infer the hydrophilicity or hydrophobicity of the material. In the present study, according the obtained regain value it is possible to presume that the fabric, which originated the textile reinforcement employed in this study, has received a treatment during its manufacture, which increased the cotton hydrophilicity, meaning also a reduction of the crystallinity and an increase of the amorphous regions on cellulose structures (Reddy and Yang 2005).

Despite of this finding, it was also possible to verify the good adhesion of all produced composites by the microscopy magnifications (32x) (Figure 4). An indicator that the proper anchor occurred between reinforcement and matrix is that neither torn fibers were observed within the polymer matrix in the places of rupture of the composites (after tested in a tester machine), nor cracks or fractures in the composite material as a whole. Figure 4 also shows the dispersion of microfibrils inside all three produced composites, being probable that they also aided to increase the good adhesion between matrices and reinforcement.

Thus, despite contradictory, it is surprising that cotton residues composed by cellulose that received a previous treatment increasing their hydrophilicity, have had such good adhesion with all assayed hydrophobic matrices (epoxy, polyester and polyurethane). Cellulose functioning as composite reinforcement generally receives a treatment for increasing its hydrophobicity, such as plasma (Sampaio2013).

It should be remarked that the present authors did not find in literature reports about the possible reasons for such contradictory behavior. A hypothesis for this good adhesion could be the tendency of cellulose molecule to form hydrogen bonds and Van der Waals forces between the various planes, including interactions between reinforcement and matrices. Besides this, as cited before, another mechanical factor could be the dispersion of microfibrils inside all three produced composites, being probable that they also aided to increase the good adhesion between matrices and reinforcement.

3.3 Sustainable design

Eco-design or sustainable design is a means to reduce or eliminate environmental impacts generated by the conventional design, replacing products and manufacturing processes for those not harmful to the environment in order to promote sustainable development. Thus, the evaluation of the Life Cycle Assessment (LCA) and product recycling are important factors to be considered. LCA aims an approach of lesser impact on the environment in all the component stages of the manufacturing process, from obtaining raw materials to the disposal of the manufactured product after consumption. LCA takes into account the product durability, its strength and its recycling.

The objective of recycling is to employ residues and reuse them in order to save raw materials and decrease the amount of accumulated waste (Ricken et al. 2008). In addition, according to Scarlato and Pontin (1992) recycling should not be the only way to obtain the raw material, however, it could be an alternative to reduce the accumulation of waste and also as a proposition for environmental education.

A fashion product has a great environmental impact, from conception to disposal. Thus, one of the ways to reduce the environmental impact would be to rethink its entire projective and productive process inserting the concepts of sustainable design, minimizing inputs and outputs during the product life cycle through eco-design (Martins and Sampaio 2006). The sustainable character must be taken into account from the pre-design to production process, generating technological changes and aiming the reduction of environmental impact (Manzini and Vezzoli, 2008).

Broega et al. (2011) suggest how eco-design methodology could be inserted in the Fashion Cycle: through a comprehensive analysis, covering all stages, eco-design is intended to integrate viable possibilities from production until discard of the products. For this purpose, the intervention of public and private institutions would also be required in order to assist with the correct planning for the collecting, management and distribution of unwanted clothing and textiles, which could then be transformed in processes of "partial or total recycling".

The studied materials could be employed in fashion design or other fields of design, especially by their visual aspects and sufficient mechanical characteristics. The present study aimed the production of new composite materials and the analysis of their tensile and aesthetical characteristics in order to collaborate also for the product eco-design methodology based on the recycling of textile wastes, which are extremely abundant in Brazil.

In spite of the employed resins not being recyclable or biodegradable, they were chosen because of the unavailability of purchasing recyclable or biodegradable resins in the Brazilian market. Furthermore, the employment of the chosen resins in this study could provide preliminary data for posterior studies employing biodegradable matrices and consequent biodegradable or recyclable fashion accessories produced from these materials.

Every project involves a creative process, which usually employs a "trial and error" method - from an existing problem possible solutions are applied. During the creative process, the individual - possessing his repertoire and personal experiences, is surrounded by a chaotic situation, subject to endless possibilities and ideas that arise when trying to find solutions to the problem (Ferrari 2011).

Salles (2008) conceptualizes the "documentation process" in the contemporary artistic creation process, basically as the repertoire of stockpiled references and experimentation that the designer uses on the development of creation. According to the same author (Salles 2012), files, documents and records appear quite recurrently in researches developed by those who dedicate themselves to criticism for understanding the processes of creation, i.e., the criticism process, which provides an approach to art from their construction procedures. Documents from creation processes (sketches, notes, audiovisual records, etc.) are taken as material records and indications of the creative route. They are secular portraits of artistic constructions, in the file creation form. Taking in account the recording issue, there are two great constants in these documents that accompany the movement of the production of artworks. They would be common features that are present in the processes under different forms. In general, these documents play two roles during the creative process: storage and experimentation. The artist finds various means for storing information, which act as auxiliaries in the course of artwork completion and nurture the artist and artwork in creation. The storage takes place in the form of notes, diaries, correspondences, etc. Generally the act of creating provisions is always present in the documentation process; however, what is stored and how it is recorded varies from case to case, even by the same artist. Another function, performed by documentation process, is to record the experimentation, making clear the inductive creation nature. In the moment of artwork execution, various kinds of hypothesis are proposed and tested. Private documents are responsible for the artwork development. They are possibilities of artworks. From this perspective, they are experimentation records always present in the creative act, found in drafts, studies, sketches, plans, scripts, models, screenplays, projects, trials, contacts, storyboards. Again, experimentation is common; the singularities arise in the principles that guide the choices. It is not possible to do any kind of generalization about the existence and the consequent use of these different material supports in the various creative processes, even in relation to the employment of such documents by a particular artist. There are variations from one artist to another and from one process to another. It was not exhausted in these listings, the possibilities of creative documentation, but what is being offered here is an overview, from which the peculiarities and individuality of each artist must be worked out case by case. In addressing this diversity, there is a point further questioned when studies about creation are presented: documents and new technologies. It has been observed that many artists employ the computer as a facilitator of their creative course and, in many cases, not at the expense of other media, which were already used, thus broadening the scope of materiality for analog records.

According to Salles (2012), we are thus faced with ever-expanding archives, seen as a kind of artist notebook, which collects things that interest the artist; it has a website and is also exhibited and published. Apparently private files are made public, whose development can be tracked. It is the storage in its ability to protect an exploration potential for future artworks; records of route (like sketches), but public. Provision and exposure are then indistinguishable.

In the present study, for the creation development of fashion accessories, besides the experimentation with the obtained composite materials, the artworks from two renowned Brazilians designers were used for the present "documentation process". Known as Campana Brothers and R. Sobral, they have employed resins or/and textile wastes to compose their works. Campana Brothers made a series of furniture called Sushi (2000 - 2004) based on the employment of different textures and types of overlapping fabrics (Campana Brothers 2012) (Figure 5a).

The designer Carlos Alberto Rezende Sobral, better known as R. Sobral, is famous for employing different types of resins in their bijoux jewelry creations, as presented in Kandinsky collection (Sobral 2012) (Figure5b). In such manner, finding that the tensile and aesthetics properties of the composite materials obtained were suitable for the production of these fashion accessories, two simple suggestions that were made for the present study are presented in Figure 6.

(a) (b)

Figure 5. Design inspirations employing textiles and resins. (a) Armchair from Sushi series of Campana Brothers (Campana Brothers 2012). **(b)** Resin necklace (Kandinsky collection) from jewelry designer R. Sobral (Sobral 2012).

(a) (b)

Figure 6. Examples of fashion design applications made of epoxy resin reinforced with textile fibers. (a) Bracelet. **(b)** Gemstone ring.

It is noteworthy that all obtained composites (from epoxy, polyester and polyurethane resins) are aesthetically similar as shown in Figure 3, as well as the molding and production technique is also similar for the three materials. Thus, for providing examples of fashion applications, taking in account only the aesthetically aspect, any of these composites could be chosen for the confection of presented accessories, since mechanical resistance is not the main aspect of this kind of application.

The composites employed in this study constitute a possibility for the textile reinforcement reduction, because besides possessing strength and cohesion functions, they could be assigned to design or fashion field for the visual attributes they provide. Moreover, they are able to supply specific needs of the consumer market, eager for news, and to furnish the demand for studies of new materials in the textile sector, which lacks specific bibliography that bring together aesthetic and technical analysis covering all production phases of materials, processes, creation and design of the final product.

At last, it is worth noting that the present study did not intend to exhaust the theme of development of fashion products from composite materials taking in account their possibilities of mechanical strength, functionality, aesthetics and recyclability or biodegradability. Hence, new trials and tests should be conducted in future studies aiming the development of new products from these composites.

4. CONCLUSIONS

The present study showed an alternative for textile recycling by means of production of thermoset composites made from thermoset resins (epoxy, polyester orthophthalic and polyurethane) reinforced with recycled cotton textile residues, on which tensile tests were carried out. The incorporation of 30% (v/v) of textile reinforcement resulted in an increase of approximately 2-fold relative to pure resin tenacity and Young's modulus for composites made from the specific epoxy resin employed in the present study; and 2-fold relative to pure resin tenacity and 3-fold more relative to Young's modulus for composites made of orthophthalic polyester resin. No significant alterations were observed for these parameters in polyurethane resin composites.

All these materials could be employed in fashion design or other fields, especially on account of their visual aspects. Despite the fact that the employed resins are not biodegradable, they were chosen because of the unavailability of purchasing recyclable or biodegradable resins in the Brazilian market. Furthermore the employment of the chosen resins in this study could provide preliminary data for posterior studies employing biodegradable matrices. The present study aimed the production of new composite materials and the analysis of their tensile characteristics in order to collaborate also for the product eco-design methodology based on recycling of textile wastes, which are extremely abundant in Brazil. In addition, the present study aimed to contribute to the scarce literature that joins aesthetic and technical analysis covering all production phases of materials, processes, creation and design of the final product.

ACKNOWLEDGEMENTS

Financial supports from CAPES (Coordination of Improvement of Higher Education Personnel – Brazilian Education Ministry), FAPESP (Sao Paulo Research Foundation) and CNPq (Brazilian National Council for Scientific and Technological Development) are gratefully acknowledged.

COMPETING INTERESTS

There are no financial or non-financial competing interests (political, personal, religious, ideological, academic, intellectual, commercial or any other) to declare in relation to this manuscript.

AUTHORS' CONTRIBUTIONS

WFZ and BMGG carried out laboratory experiments, documentation process of design and artwork, proposition of fashion accessories and drafted part of the manuscript. WD and JBR carried out analyses of experimental results (in technical, statistical, design and artistic aspects), drafted part of the manuscript, reviewed critically the previous and current versions of manuscript, have given final approval of the version to be published and agree to be accountable for all aspects of the study in ensuring that questions related to the accuracy or integrity of any part of the study are appropriately investigated and resolved. All authors read and approved the final manuscript.

REFERENCES

1. ABNT, ISO 139:2005 – Brazilian Association of Technical Standards and International Organization for Standardization. (2005). *Têxteis - Atmosferas-padrão para condicionamento e ensaio ("Textiles - Standard atmospheres for conditioning and testing")*. São Paulo, SP, Brazil: ABNT.

2. ABNT, NBR 13041:2004 - Brazilian Association of Technical Standards - Brazilian Standard. (2004). *Nãotecido - Determinação da resistência à tração e alongamento - Método de tiras e Grab ("Nonwoven - Determination of tensile strength and elongation – Strip and Grab Methods")*. São Paulo, SP, Brazil: ABNT.

3. Adanur, S. (2001). *Handbook of Weaving*. Lancaster, PA, USA: Technomic.

4. Almeida, M, & Rezende, M. (2001). Interlaminar shear strength of composites with epoxy resin with different arrangement of fibers in the presence of voids. *Polímeros, 1*(4), 182–189.

5. Alwood, J, Laursen, SE, Rodríguez, CM, & Bocken, NMP. (2006). *Well dressed? The present and future sustainability of clothing and textiles in the United Kingdom*. London, UK: University of Cambridge - Institute for Manufacturing.

6. Amico, SC. (2010). Vegetable fibers as multifunctional materials. *Matéria (Rio de Janeiro), 15*(2), 355–363.

7. Basra, AS, & Malik, CP. (1984). Development of the cotton fiber. *International Review of Cytology, 89*, 65–113.

8. Bergamasco, C. (2012). *Jenswear volta com tudo nessa temporada ("Jeanswear come back up in this season")*: 17 April 2012. http://gofw.com.br/vestuario/janswear-volta-com-tudo-nessa-temporada. Accessed 12 June 2014.

9. Broega, AC, Carvalho, C, & Moraes, C. (2011). Methodology of eco-design in the fashion cycle: reuse and recycling of wasted clothing. In *Proceedings of the VI International Congress of Design Research (CIPED), Oct. 10-12, 2011*. Lisbon, Portugal: Calouste Goulbekian Foundation.

10. Callister, WD, Jr. (2007). *Materials Science and Engineering* (7th ed.). New York, USA: John Wiley & Sons, Inc.

11. Campana Brothers. (2012). *Fernando and Humberto Campana - Sushi collection - armchair*:http://campanas.com.br/. Accessed 03 December 2012.

12. Carneiro, MRD. (2013). *Plasma treatment of coconut fibers for obtaining nanofibers and producing phb matrix composites* (Graduation final paper in Materials Engineering). Rio de Janeiro (RJ) – Brazil: Polytechnical School. National University of Rio de Janeiro (UFRJ).

13. Carvalho, FR. (2005). *Composites of sisal fibers for use in reinforcement of wooden structures. PhD Thesis in Material Science and Engineering*. São Carlos: Polytechnical School, Institutes of Physics and Checmistry. University of São Paulo (USP).

14. CNI – Brazilian National Confederation of Industries, & ABIT – Brazilian Association of Textile and Apparel Industry. (2012).

Têxtil e confecção: inovar, desenvolver e sustentar ("Textile and clothing: innovate, develop and sustain"). Brasília, DF, Brazil: CNI.

15. Costa, ACR, & Rocha, ERP. (2009). Panorama da cadeia produtiva têxtil e de confecções e a questão da inovação ("Overview of the textile and apparel supply chain and the question of innovation"). *BNDES Setorial, 29*, 159–202.
16. Fabriclink – Fabric University. (2012). *Yarn facts: ring spinning*: http://www.fabriclink.com/university/yarnfacts.cfm. Accessed 08 December 2012.
17. Ferrante, E. (2011). *Biotechnology applied to textile industry*: http://textileindustry.ning.com/profiles/blogs/biotecnologia-aplicada-a-6. Accessed 05 December 2012.
18. Ferrari, DOA. (2011). *The comparative study between the criative process in the architechture and in the jewellery emphasizing the creations of Frank Gehry* (Master Science Dissertation in Architecture and Urbanism). São Paulo (SP) – Brazil: College of Architecture and Urbanism of University of São Paulo.
19. Furlan, LG, Duarte, UL, & Mauler, RS. (2012). Evaluation of oat-reinforced polypropylene composite properties.*Química Nova, 35*(8), 1499–1501.
20. Gil, L. (2009). Cork composites: a review. *Materials, 2*, 776–789.
21. ISO/TR 6741–4–1987 - International Standards Organization / Technical Report. (1987). *Textiles - Fibres and yarns - Determination of commercial mass of consignments- Part 4: Values used for the commercial allowances and the commercial moisture regains*. Switzerland, Geneva: ISO.
22. Kaswell, ER. (1963). *Wellington Sears Handbook of Industrial Textiles*. Cambridge, USA: Massachusetts Institute of Technology (MIT) and Wellington Sears Company.

23. Kolpak, FJ, Weih, M, & Blackwell, J. (1978). Mercerization of cellulose: 1. Determination of the structure of mercerized cotton. *Polymer, 19*(2), 123–131.

24. Leite, F. (2011). *Bom Retiro Quer Limpeza - Funcionários das Centenas de Pequenas Confecções da Região Depositam nas Ruas Sacos com Retalhos de Tecido (Bom Retiro wants cleanliness - workers of hundreds of small confections in this region are depositing in the streets bags with fabric scraps).* : . 29 September 2011. http:// textileindustry.ning.com/forum/topics/bom-retiro-quer-limpeza. Accessed 4 Jan 2012.

25. Manzini, E, & Vezzoli, C. (2008). *O desenvolvimento de produtos sustentáveis: os requisitos ambientais dos produtos industriais ("The development of sustainable products: environmental requirements of industrial products")* (1st ed.). São Paulo, SP, Brazil: EDUSP.

26. Marsh, G, Trynka, P, & Marsh, J. (2003). *Denim: l'épopée illustrée d'un tissu de legend ("Denim: illustrated epic of a legendary fabric").* Paris, France: Editions Du Collectioneur.

27. Martins, SB, & Sampaio, CP. (2006). Ecodesign e design sustentável: proposta de método para um workshop ("Ecodesign and sustainable design: a methodology proposal for a workshop"). In *Proceedings of the 7th Brazilian Congress of Research and Development in Design (P&D), Aug. 7-9, 2006*. Curitiba, PR, Brazil: Centro Universitário Positivo (Unicenp).

28. MDIC - Brazilian Ministry of Development, Industry and External Trade, & ABIT - Brazilian Association of Textile and Apparel Industry. (2011). *MDIC e ABIT realizam pesquisa inédita sobre perfil do consumidor de vestuário ("MDIC and ABIT conduct original research about apparel consumer profile")*: 08 November 2011.http://www.mdic.gov.br/sitio/interna/noticia.php?area=1& noticia=11095. Accessed 13 January 2014.

29. Mesdan-Lab. (2005). *Manuale d'uso e manutenzione ("Operating instructions and maintenance"): Video Analyser 2000, Microlab Advance, Macrolab Advanced* (1st ed.). Raffa de Puegnago, BS, Italy: Mesdan S.p.A.

30. Nascimento, WCA, & Martins, MLL. (2006). Production of proteases by Bacillus sp. SMIA-2 grow on whey and corn steep liquor and compatibility of the enzyme with commercial detergents. *Food Science and Technology (Campinas), 26*(3), 582–588. (2006) .doi: 10.1590/S0101-20612006000300016.

31. Poliresinas. (2012). *Resinas poliéster insaturadas ("Unsaturated polyester resins")*. Poliresinas Indústria e Comércio de Resinas Ltda. São Bernardo do Campo/SP, Brazil. http://www.poliresinas.com.br/resinas-poliester-insaturadas.html. Accessed 09 December 2012.

32. Queiroz, LL. (2003). *Urban wastes and the issue of ecodesign: the relevance of biodegradable plastic*(Master Science Dissertation in Design). Rio de Janeiro (RJ) – Brazil: Pontifical Catholic University of Rio de Janeiro.

33. Raqueza, JM, Deléglisea, M, Lacrampea, MF, & Krawczaka, P. (2010). Thermosetting (bio) materials derived from renewable resources: A critical review. *Progress in Polymer Science, 35*(4), 487–509.

34. Reddy, N, & Yang, Y. (2005). Properties and potential applications of natural cellulose fibers from cornhusks.*Green Chemistry, 7*, 190–195.

35. Ricken, LC, Pozza, PP, & Teixeira, JA. (2008). Ecodesign: escolha de materiais visando sustentabilidade e processos de fabricação ("Ecodesign: choice of materials aiming sustainability and manufacturing processes"). In *Proceedings of the 8th Brazilian Congress of Research and Development in Design (P&D), Oct. 8-11,*

2008. São Paulo, SP, Brazil: Associação de Ensino e Pesquisa de Nível Superior de Design do Brasil.

36. Rosario, F, Pachekoski, WM, Santos, SF, Savastano, H, Jr, & Casarin, SA. (2011). Virgin and recycled polypropylene composites reinforced with Sisal by-product. *Polímeros, 21*(2), 90–97.

37. Salles, CA. (2008). *Crítica Genética: fundamentos dos estudos genéticos sobre o processo de criação artística ("Genetic Criticism: Foundations of genetic studies about the process of artistic creation")* (3rd ed.). São Paulo, SP, Brazil: EDUC.

38. Salles, CA. (2012). Arquivos nos processos de criação contemporâneos ("Files in contemporary creation processes"). In *Proceedings of the National Association of Researchers in Plastic Arts, Sep. 24-29, 2012*. Rio de Janeiro, RJ, Brazil: Associação Nacional de Pesquisadores em Artes Plásticas.

39. Sampaio, JA. (2013). *Characterization of plasma treated cellulose fibers* (Graduation final paper in Materials Engineering). Rio de Janeiro (RJ) – Brazil: Polytechnical School. National University of Rio de Janeiro (UFRJ).

40. Santana Textiles. (2014). *Loco serious denim – collection rigid – Absynto*: http://www.santanatextiles.com/locofiltro.php?t=5. Accessed 12 June 2014.

41. Santanaense. (2014). *Glossário - Mercerização ("Glossary - Mercerization")* http://www.santanense.com.br/sportswear/glossario/1/M. Accessed 12 June 2014.

42. Saville, BP. (2007). *Physical testing of textiles*. Cambridge, UK: The Textile Institute and Woodhead Publishing Limited.

43. Scarlato, FC, & Pontin, JA. (1992). *Do nicho ao lixo: sociedade, ambiente e educação ("From the cradle to grave: society,*

environmental and education") (18th ed.). São Paulo, SP, Brazil: Atual Editor.

44. SINDITEXTIL - Union of Spinning and Weaving Industries of the State of São Paulo – Brazil. (2012). *Retalho Fashion é destaque de evento na Rio + 20 ("Retail Fashion is featured event at Rio +20")* 20 June 2012.http://www.sinditextilsp.org.br/index.php/materias/item/841-retalho-fashion-%C3%A9-destaque-de-evento-na-rio%20 20. Accessed 13 March 2014.

45. Sobral, R. (2012). *Kandinsky collection - necklace.* http://en.sobraldesign.com.br/#/undefined/Collections. Accessed 03 December 2012.

46. UTL - Technical University of Lisbon. (2012). *Propriedades Mecânicas de Metais, Cerâmicos e Polímeros ("Mechanical Properties of Metals, Ceramics and Polymers")* http://disciplinas.ist.utl.pt/qgeral/mecanica/MetCerPol.pdf Accessed 11 November 2012.

47. Vianna, F. (2011). *Personal interview of Eng. Francisco Vianna from LOGA – Environmental Logistics of São Paulo S. A.* http://www.loga.com.br/. 08 August 2011.

48. Wadaa, M, Ikeb, M, & Tokuyasub, K. (2010). Enzymatic hydrolysis of cellulose I is greatly accelerated via its conversion to the cellulose II hydrate form. *Polymer Degradation and Stability, 95*(4), 543–548.

49. Zia, KM, Bhatti, HN, & Bhatti, IA. (2007). Methods for polyurethane and polyurethane composites, recycling and recovery: a review. *Reactive and Functional Polymers, 67*(8), 675–692.

11

APPLICATION OF *TERMINALIA CHEBULA* NATURAL DYE ON WOOL FIBER—EVALUATION OF COLOR AND FASTNESS PROPERTIES

Mohd Shabbir[1] , Shahid Ul Islam[1] , Mohd Nadeem Bukhari[1] , Luqman Jameel Rather[1] , Mohd Ali Khan[2] and Faqeer Mohammad1*

[1]Department of Chemistry, Jamia Millia Islamia, New Delhi 110025, India.

[2] Department of Post Harvest Engineering and Technology, Faculty of Agricultural Sciences, AMU, Aligarh, UP 202002, India.

ABSTRACT

In the present study, *Terminalia chebula* (Myrobalan/Harda) natural dye extract was used for the development of eco-friendly shades on woolen yarn with different hues and tones. The effect of dye concentration on color strength (*K*/*S*) of woolen yarn dyed with *T. chebula* was assessed. Increasing the concentration of dye decreased lightness (*L**) values of woolen yarn samples, indicating darker shades. Different metal salts such as alum, ferrous sulfate, and stannous chloride were used to enhance the fastness properties (light, wash, dry and wet rubs) of dyed woolen yarn. Pre-treatment of woolen yarn samples with metal salts has shown encouraging results with better fastness properties and enhanced color strength values. Five percent ferrous sulfate mordanted samples show greater saturation with increasing dye concentration from 0.5 to 15 % (o.w.f.). Fourier-transform infrared (FT-IR) analysis of *T. chebula* dye extract shows presence of carbonyl and hydroxyl functionalities.

KEYWORDS

Natural dye, Harda Mordants, Color strength, Fastness properties

1. BACKGROUND

Natural dyes have been used since time immemorial to color different textile materials. In the nineteenth century, synthetic dyes in view of their low cost, large variety of shades, and high dyeability overtook the use of natural dyes (Samantha and Agarwal 2009; Yusuf et al. 2013). Synthetic dyes and particularly azo-based dyes are recently discovered to pollute the environment. This has motivated textile researchers to reintroduce natural colorants from renewable resources once again into the dyeing industry. Natural dyes produced from plants, animals, insects, and minerals are mostly non-toxic, non-carcinogenic, biodegradable, and environment-friendly in nature with some biological activities (Khan et al. 2010; Shahid et al. 2013). To achieve the sophisticated demand of modern people, a lot of research has been undertaken in the field of natural dyes for obtaining colorful shades on textiles and evaluation of their tolerance to the light, wash, and rub effects (Khan et al. 2015; Islam and Mohammad 2015). Metallic mordants as well as bio-mordants can be used to enhance the color characteristics and fastness properties of natural dyes on a textile substrate (Dalby 1993; Vankar et al. 2008). It is also reported that natural dyes can not only be used to get natural shades but they can also provide functionalities to fabrics such as antibacterial activity (M. I. Khan et al. 2011; S. A. Khan et al. 2012; Shahid et al. 2012), antifungal activity (Yusuf et al. 2015), and ultraviolet protection (Zhou et al. 2015).

Tannin-based dyes are used for textile dyeing for their good substantivity towards them because of more numbers of auxochromic groups in coloring components. *Terminalia chebula* is a medicinal plant grown in India and Southeast Asia (Khan et al. 2005). It is one of the constituents of triphala used in India as medicine. Fruits of *T. chebula* are included in the Indian pharmacopeia under the category astringent. It possesses laxative, diuretic, cardiotonic, and hypoglycemic properties (Naik et al. 2004). The major phytoconstituents present in the fruits are hydrolysable tannins, gallic acid, chebulic acid, chebulic ellagitannins, and gallate esters (Pfundstein et al. 2010).

The present research paper is aimed to explore the dyeing potential of *T. chebula* on wool fiber. The effect of the dye concentration and different types of metal mordants such as alum, ferrous sulfate, and stannous chloride on color characteristics of developed shades and their fastness properties are also studied.

1.1 Materials

Wool yarn (100 % semi-worsted 60 counts) was purchased from MAMB Woollens Ltd. Bhadohi, UP, India. *T. chebula* dye extract in powder form was purchased from Sir Biotech India Ltd. Kanpur, UP, India. Metallic mordants potash alum ($Al_2K_2(SO_4)_4 \cdot 24H_2O$), iron sulfate ($FeSO_4 \cdot 7H_2O$), and stannous chloride ($SnCl_2 \cdot 2H_2O$) were used of laboratory grade.

1.2 Instrumentation

Fourier-transform infrared (FT-IR) spectra of *T. chebula* dye powder were obtained on a Perkin Elmer Spectrum RXI FT-IR system in order to investigate and observe auxochromic groups participating in wool-dye interactions (with the resolution of 4 cm^{-1}). Bands in the FT-IR spectra were resolved in accordance with literature data. A Perkin Elmer Lambda-40 double beam UV-visible spectrophotometer was employed for recording absorbance values of dye solutions. A pH/mV meter (BD

1011) from Decibel digital technologies was used for measuring pH of dye solutions.

2. METHODS

2.1 Mordanting

Pre-mordanting method was opted for which three eco-friendly mordants alum ($Al_2K_2(SO_4)_4 \cdot 24H_2O$), iron sulfate ($FeSO_4 \cdot 7H_2O$), and tin chloride ($SnCl_2.2H_2O$) were selected. Woolen yarn samples were soaked in water before mordanting. The mordants were dissolved in water, and soaked woolen yarns were immersed into mordant solution at about 30 °C. Temperature of mordant solution was raised at a constant rate up to 91–93 °C and kept at this temperature for 60 min with constant stirring. Unused mordants on woolen yarn were removed by rinsing with tap water.

2.2 Optimization of mordants

Three different mordants such as alum ($Al_2K_2(SO_4)_4 \cdot 24H_2O$), iron sulfate ($FeSO_4 \cdot 7H_2O$), and tin chloride ($SnCl_2 \cdot 2H_2O$) used in this study were optimized for their better performance on wool in terms of achieving higher color strength and better fastness properties without sacrificing much of the mordant. Alum, iron sulfate, and tin chloride were optimized in the concentration range of 1.0–10, 1.0–5, and 0.1–1 % o.w.f (on weight of fiber). In view of the toxic nature of some metal salts, higher concentrations were not selected in this study.

Optimization was done on the basis of absorbance values (percentage exhaustion), recorded before and after dying with a UV-vis spectrophotometer, and percentage exhaustion was calculated by using the following equation.

$$\text{Percentage exhaustion} = \frac{\text{initial absorbance} - \text{final absorbance}}{\text{initial absorbance}} \times 100 \qquad (1)$$

Based on the absorption and visual appearance, each mordant was selected in a particular optimized concentration (Prabhavathi et al. 2014) and was evaluated for colorimetric and fastness characteristics.

2.3 Dyeing

Un-mordanted and mordanted woolen yarns were immersed into dyebaths of varying concentrations (15, 8, 3, 1, and 0.5 % o.w.f.) maintained at M:L (material to liquor) ratio of 1:40 at neutral pH conditions. Temperature of dyebath was raised at a constant rate up to 91–93 °C and kept at that temperature for 60 min with constant stirring to achieve uniform dyeing. Acid (HCl) (pH = 4.0) and alkali (Na_2CO_3) (pH = 9.0) after-treatment was performed to dyed samples to demonstrate the effect of pH on dyeing processes. Dyed and after-treated woolen yarn samples were washed with non-ionic detergent Safewash, Wipro (5 ml/L), and rinsed with tap water. The fibers were dried in shade at room temperature.

2.4 Color measurements

Color measurements of dyed woolen yarn were carried out by following standard procedures. Estimation of color parameter values in terms of K/S and CIE-$L^*a^*b^*$ values were recorded on Gretag Macbeth color-eye 7000A spectrophotometer connected to a computer with installed software of MiniScan XE Plus. Color strength (K/S) value was calculated by using Kubelka-Munk equation.

$$\frac{K}{S} = \frac{(1-R)^2}{2R} \qquad (2)$$

where K is the absorption coefficient, S is the scattering coefficient, and R is the reflectance of dyed samples.

Chroma (c^*) and hue angles ($h°$) were calculated using the following equations:

$$\text{Chroma } (c^*) = \sqrt{a^2 + b^2} \tag{3}$$

$$\text{Hue angle } (h^o) = \tan^{-1}(^b/_a) \tag{4}$$

2.5 Fastness determination

Light fastness of the dyed samples was conducted on Digi light Nx™, having water cooled by Mercury Blended Tungsten lamp, according to test method ISO 105-B02:1994 (Amd.2:2000). Wash fastness was measured in Digi wash SS™ (Laundrometer) as per the ISO 105-C06:1994 (2010) specifications. Dry and wet rub fastness of dyed samples were tested using Digi crock™ (Crockmeter) as per Indian standard IS 766:1988 (reaffirmed 2004) based on ISO 105-X12:2001 by mounting the fabric on a panel and giving ten strokes for both dry and wet rub fastness tests. The samples were assessed for staining on white adjacent fabric (cotton and wool).

3. RESULTS AND DISCUSSION

3.1 FT-IR spectra of Terminalia chebula dye

FT-IR analysis was used to identify the possible auxochromes responsible for the substantivity of dye. FT-IR spectra (Fig. 1) of extract powder of *T. chebula* were recorded which showed bands at 3240 and 2980 cm^{-1}due to O–H and aromatic C–H stretching vibrations, respectively; presence of bands at 1710 and 1595 $cm^{-1}$is due to C=O and C=C stretching vibrations, respectively. Stretching vibrations at 1201 and 1040 cm^{-1} are for C–O stretching peaks.

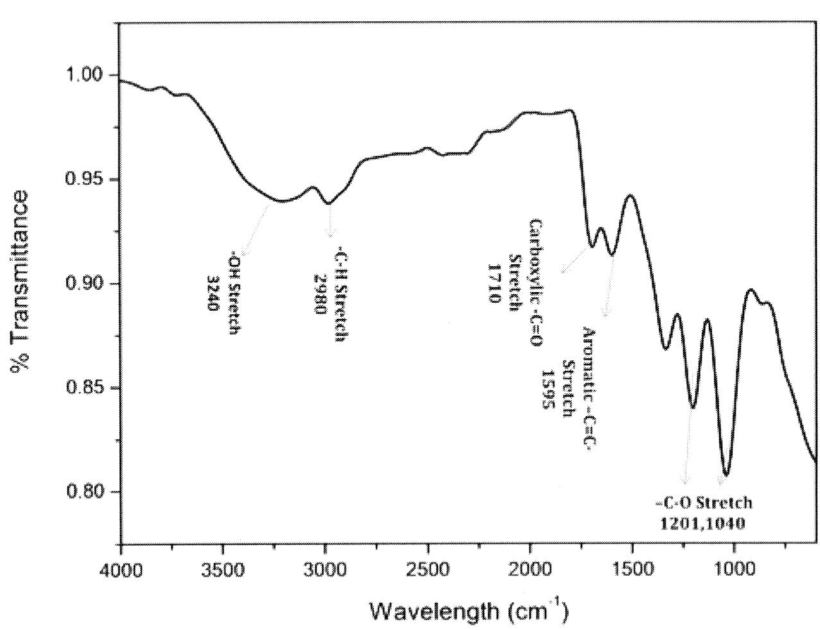

Figure 1. FT-IR spectrum of *Terminalia chebula* dye

3.2 Optimization of mordant concentrations

Metal salt mordants have different interactions with wool and thereby may darken, brighten, or drastically alter the final color of the dyed wool samples. In this study, wool yarns were pre-treated with different types of metal salts (alum, ferrous sulfate, stannous chloride).

Figure 2a–c shows the effect of alum, iron sulfate, and tin chloride concentrations on the percentage exhaustion values of *T. chebula*-dyed woolen yarns. On increasing the concentration of alum (1–10 % o.w.f.), iron sulfate (1–5 % o.w.f.), and tin chloride (0.1–1.0 % o.w.f.) mordants, respectively, percentage exhaustion values increase. This increase may be attributed to the increasing interaction of respective metal ions or their higher coordinating ability with wool fiber and dye molecules.

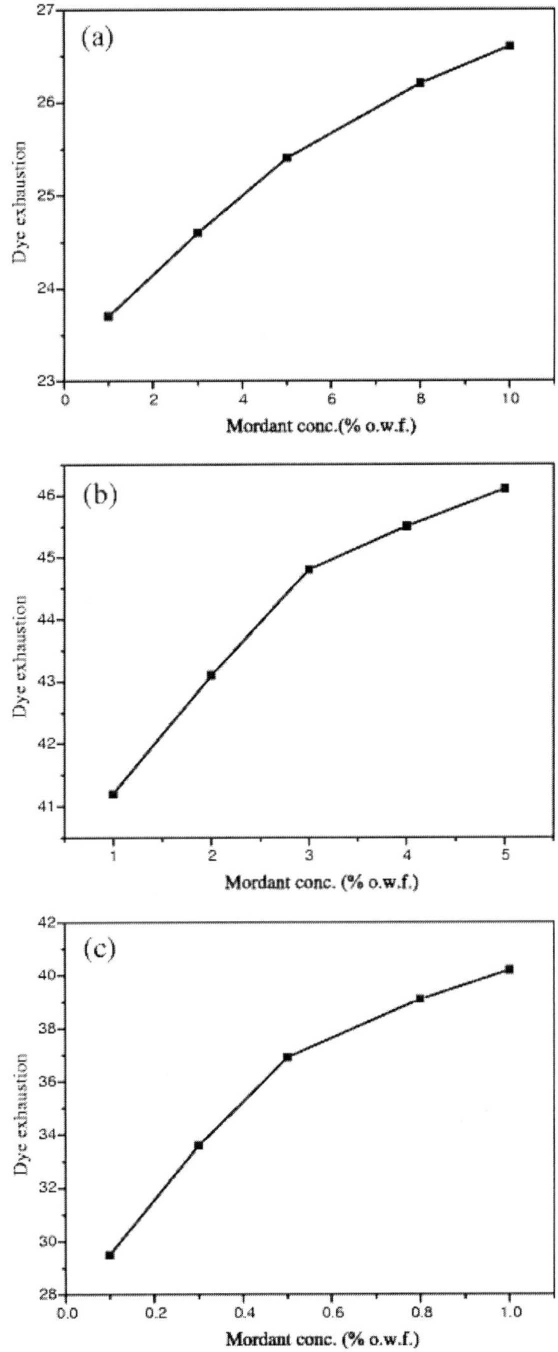

Figure 2. Optimization of mordants concentration. **a** Alum. **b** Iron. **c** Tin

From the experimental results, 10.0 % (o.w.f.) alum, 5.0 % (o.w.f.) iron sulfate, and 1.0 % (o.w.f.) tin chloride concentrations were found to give maximum exhaustion results and were taken as optimized mordant concentrations for subsequent dyeing experiments with *T. chebula* natural dye.

3.3 Color characteristics

Colorimetric characteristics (K/S and CIE-$L^*a^*b^*$ values) of woolen yarn dyed with different concentrations of*T. chebula* (15.0, 8.0, 3.0, 1.0 and 0.5 % o.w.f.) with optimized mordant concentrations were analyzed (Fig. 3). Negatively charged dye molecules (anions) and metal (cations) have strong affinity for positively charged amino and negatively charged carboxyl groups (isoelectric nature of wool fiber), respectively. Electrostatic forces of attraction (ionic bonding) between coloring components and wool functional groups are responsible for the uptake of dye by wool in dyeing bath in addition to hydrophobic forces of attraction (van der Waals forces). From the dye-metal coordination complexes, the vacant sites are used to form coordinate bonds with the uncharged amino ($-NH^-$) and carbonyl (C=O) groups of the amide group of wool (Fig. 4).

3.4 Effect of concentration of dye on color strength of dyed woolen yarn

With the increase in the concentration of dye from 0.5 to 15.0 % (o.w.f.), an increase in color strength (K/S) is observed resulting in deeper shades with increase in the dye adsorption. Due to increase in concentration of dye in dyebath, concentration gradient increases, which eventually results in an increase in the rate of diffusion of dye onto the surface of woolen yarn from the dyebath solution. A lower concentration of dye produces shades of gray-colored tones while a higher concentration of dye shifts color to the yellow region.

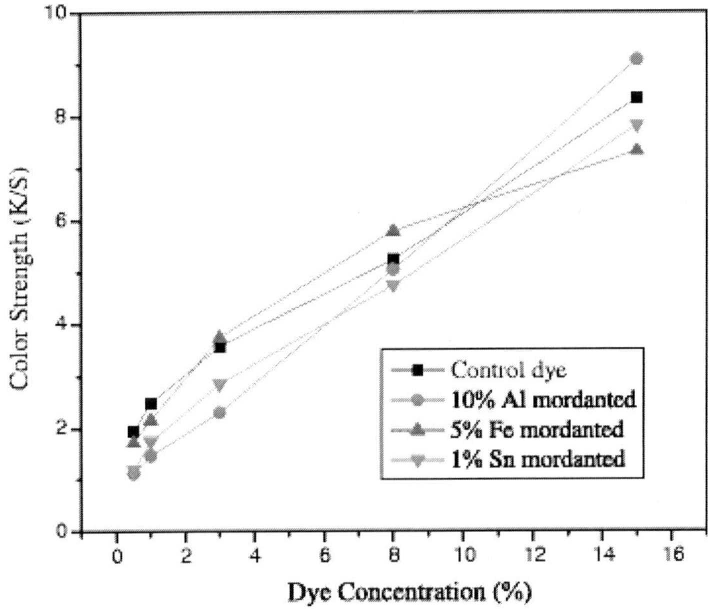

Figure 3. Effect of dye concentration on color strength (*K*/*S*) of dyed woolen fiber

3.5 Effect of pH

Wool yarns were dyed with *T. chebula* in a dyebath maintained at neutral pH. To widen the shade range, dyed wool yarns were after-treated with acidic and alkaline solutions. After treatment of dyed wool yarn samples with acidic and alkaline media, appreciable changes in color parameters and fastness properties were observed. Acidic treatment lightens the shades as supported by low color strength values which may be due to hydrolysis of complexes and wool-dye interaction in acidic medium. Alkaline medium treatment improved the wash fastness (color change), but acidic medium treatment decreases the tolerance to washing. The different colors obtained on wool in neutral, acidic, and alkaline media are shown in Table 5.

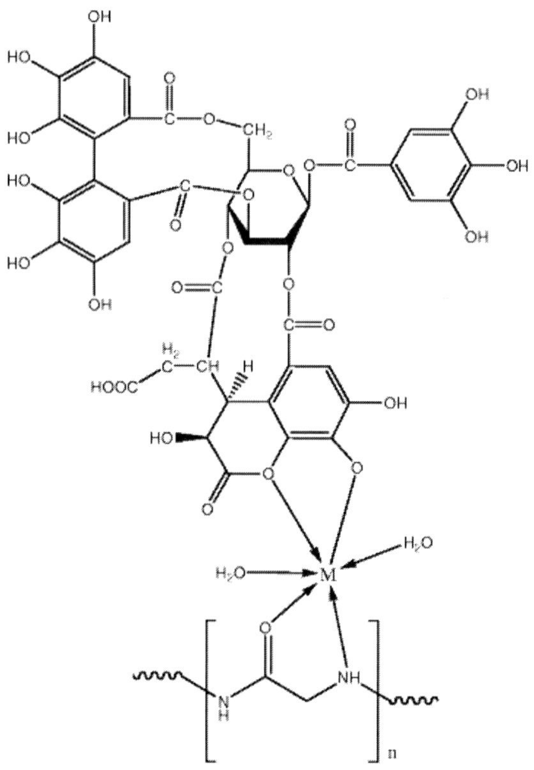

Figure 4. Complex formation between chebulagic acid, metal ion, and functional groups of wool fiber

3.6 Effects of mordants on colorimetric characteristics

Some transition metal ions and particularly iron salts in our case strongly bind with natural dye molecules due to their ability to form strong coordination complexes and thus produce deep color on the fabric (Uddin2014). Better octahedral complexation property of iron leads to higher *K*/*S* values and the color change to grayish of iron-mordanted dyed woolen yarn than un-mordanted yarn (Mihalick and Donnelly 2006). Metal mordants highly affect the lightness and a^* and b^* values (Tables 1, 2, 3, and 4); as in the case of alum and tin, lightness increases up to 75 or more; and the a^*-b^* plot (Figs. 5 and 6) shifts towards the more yellow region owing to the lightening of shade property of alum and tin mordants. In the case of iron mordants, lightness decreases up to 60 or less, and the a^*-b^* plot (Fig. 7) shifts far away from the yellow coordinate in comparison to control-dyed (Fig. 8) woolen yarn owing to the saddening property of iron mordants. The chroma values (c^*) were found minimum in the case of iron-mordanted samples in the range of 11–20 and maximum in alum-mordanted samples in the range of 18–24, whereas the values of tin-mordanted samples were found in the range of 18–22. The hue angles are in the range of 72° to 87°, and all the dyed woolen yarns were found in the yellow-red quadrant. Shade cards of all dyed samples are given in Table 5.

3.7 Fastness properties

Light, wash, and rub fastness properties (Tables 1, 2, 3, and 4) were evaluated on a grayscale and observed as average to excellent range.

3.8 Light fastness

For the light fastness, it was observed that all the woolen yarns dyed with *T. chebula* showed very good fastness results to light. The alum-mordanted dyed woolen yarns were found comparatively less tolerable towards light in comparison to control-, iron-, or tin-mordanted dyed woolen yarns (Maulik and Agarwal2014). This is probably due to weak coordination bonding of alum with dye molecule, and the photolytic degradation happens, but in the case of iron, strong coordination bonding prevents photolytic degradation by protecting the chromophore with dissipation of energy of absorbed photons to the chelate structure formed with metal ions (Jothi 2008).

Table 1. CIE-$L^*a^*b^*$ values and fastness properties of un-mordanted dyed wool samples

								Wash fastness	Rub fastness		Light fastness
Dye %	Medium (pH)	L^*	a^*	b^*	c^*	$h°$	K/S	c.c.	Dry	Wet	
15	Neutral	47.6	4.8	17.8	18.4	74.8	8.33	3	4/5	4/5	5
	Acidic	49.9	5.5	18.6	19.4	73.5	6.95	4	4/5	3/4	5
	Alkaline	43.4	6.4	19.4	20.4	71.9	15.27	5	4/5	4/5	5
8	Neutral	48.1	4.1	16.1	16.6	75.8	5.23	5	4/5	4/5	5
	Acidic	51.4	5.9	18.6	19.5	72.4	5.71	4	4/5	4/5	5
	Alkaline	49.5	5.6	19.6	20.4	73.9	7.42	2	4/5	4/5	5
3	Neutral	55.4	4.4	17.9	18.4	76.3	3.56	4	4/5	4/5	5
	Acidic	57.0	4.4	18.6	19.2	76.6	3.80	4	4/5	4/5	5
	Alkaline	56.4	4.3	18.5	19.0	76.8	3.39	3	4/5	4/5	5
1	Neutral	57.1	3.9	15.9	16.4	76.2	2.48	4/5	4/5	4/5	5
	Acidic	62.0	4.7	19.7	20.2	76.6	2.56	3	4/5	4/5	5
	Alkaline	59.7	3.6	16.5	16.9	77.6	2.24	3/4	4/5	4/5	5
0.5	Neutral	60.8	2.8	15.3	15.6	79.6	1.94	4	4/5	4/5	5
	Acidic	65.9	4.2	19.9	20.3	78.1	1.99	3/4	4/5	4/5	5
	Alkaline	63.4	2.8	15.4	15.7	79.6	1.59	4/5	4/5	4/5	5

Abbreviations: c.c. color change

Table 2. CIE-$L^*a^*b^*$ values and fastness properties of alum-mordanted dyed wool samples

Dye %	Medium (pH)	L^*	a^*	b^*	c^*	$h°$	K/S	Wash fastness	Rub fastness		Light fastness
								c.c.	Dry	Wet	
15	Neutral	54.2	2.6	22.3	22.5	83.3	9.07	4	4	3	4/5
	Acidic	55.8	5.0	19.5	20.1	75.5	4.63	3	4/5	4	4/5
	Alkaline	51.8	4.7	22.3	22.8	77.9	10.43	4	4	3	4/5
8	Neutral	61.9	2.1	23.4	23.5	84.7	5.03	4	4	4	4/5
	Acidic	56.4	4.6	17.9	18.5	75.5	3.65	3/4	4/5	4/5	4/5
	Alkaline	62.6	3.4	23.3	23.5	81.4	4.04	4	4/5	4/5	4/5
3	Neutral	69.6	1.6	22.5	22.6	85.9	2.29	4	4/5	4/5	4/5
	Acidic	62.6	3.4	17.9	18.2	79.0	2.26	4	4/5	4/5	4/5
	Alkaline	70.7	2.5	21.2	21.3	83.2	1.76	3/4	4/5	4/5	4/5
1	Neutral	73.2	1.3	20.6	20.7	86.1	1.46	4	4/5	4/5	4/5
	Acidic	66.8	3.2	18.7	18.9	80.2	1.95	4	4/5	4/5	4/5
	Alkaline	72.9	2.1	20.7	20.8	84.1	1.35	4	4/5	4/5	4/5
0.5	Neutral	73.8	1.3	18.7	18.8	85.9	1.12	4	4/5	4/5	4/5
	Acidic	75.0	1.9	18.2	18.3	83.8	1.02	2	4/5	4/5	4/5
	Alkaline	72.9	2.3	20.1	20.2	83.3	1.25	4	4/5	4/5	4/5

Abbreviations: c.c. color change

Table 3. CIE-$L^*a^*b^*$ values and fastness properties of iron-mordanted dyed wool samples

Dye %	Medium (pH)	L^*	a^*	b^*	C^*	$h°$	K/S	Wash fastness	Rub fastness		Light fastness
								c.c.	Dry	Wet	
15	Neutral	47.3	4.1	16.2	16.7	75.6	7.32	4	4/5	4/5	5
	Acidic	47.4	3.8	16.3	16.7	76.8	7.28	4	4/5	4/5	5
	Alkaline	38.2	6.2	17.7	18.8	70.5	21.15	3/4	4/5	4	5
8	Neutral	51.9	4.5	17.8	18.4	75.6	5.78	3/4	4/5	4/5	5
	Acidic	52.9	4.9	18.8	19.4	75.3	5.48	3	4/5	4/5	5
	Alkaline	42.9	6.1	19.2	20.2	72.3	15.57	4/5	4/5	4/5	5
3	Neutral	54.9	3.6	16.1	16.5	77.3	3.73	4	4/5	4/5	5
	Acidic	56.3	3.3	16.1	16.5	78.3	3.17	4	4/5	4/5	5
	Alkaline	47.6	5.2	19.6	20.3	75.0	9.57	4	4/5	4/5	5
1	Neutral	55.9	2.0	11.5	11.7	79.9	2.15	4	4/5	4/5	5
	Acidic	67.3	3.3	16.8	17.2	78.8	1.42	3/4	4/5	4/5	5
	Alkaline	55.1	3.6	14.9	15.3	76.3	2.67	3/4	4/5	4/5	5
0.5	Neutral	57.5	1.2	10.3	10.4	83.2	1.72	3/4	4/5	4/5	5
	Acidic	67.5	2.4	15.7	15.9	81.3	1.32	2	4/5	4/5	5
	Alkaline	56.3	2.9	12.6	13.0	76.7	2.12	4	4/5	4/5	5

Abbreviations: c.c. color change

Table 4. CIE-$L^*a^*b^*$ values and fastness properties of tin-mordanted dyed wool

Dye %	Medium (pH)	L^*	a^*	b^*	C^*	$h°$	K/S	Wash fastness	Rub fastness		Light fastness
								c.c.	Dry	Wet	
15	Neutral	54.8	4.9	20.6	21.2	76.4	7.81	4	4/5	4	5
	Acidic	53.9	5.6	20.3	21.1	74.5	8.20	3/4	4/5	4	5
	Alkaline	46.2	4.2	19.2	19.7	77.5	12.98	4	4/5	4	4/5
8	Neutral	62.4	4.0	21.5	21.8	79.3	4.73	4	4/5	4	5
	Acidic	61.7	4.6	21.2	21.7	77.6	4.52	3/4	4/5	4	4/5
	Alkaline	54.8	4.8	21.6	22.1	77.3	7.30	4	4/5	4	5
3	Neutral	62.5	3.0	18.0	18.2	80.4	2.84	4/5	4/5	4	5
	Acidic	62.9	3.6	19.4	19.7	79.3	3.33	3/4	4/5	4	5
	Alkaline	59.7	3.3	20.2	20.5	80.5	4.13	4	4/5	4	5
1	Neutral	70.4	1.7	18.6	18.7	84.8	1.74	4/5	4/5	4/5	5
	Acidic	69.7	2.6	18.5	18.7	81.8	1.81	3/4	4/5	4/5	5
	Alkaline	68.1	2.3	20.1	20.3	83.3	2.17	4	4/5	4/5	5
0.5	Neutral	75.1	1.1	18.6	18.7	86.7	1.19	4/5	4/5	4/5	5
	Acidic	75.0	1.8	18.6	18.7	84.4	1.22	4	4/5	4/5	5
	Alkaline	73.8	1.4	20.4	20.5	86.0	1.41	4	4/5	4/5	5

Abbreviations: c.c. color change

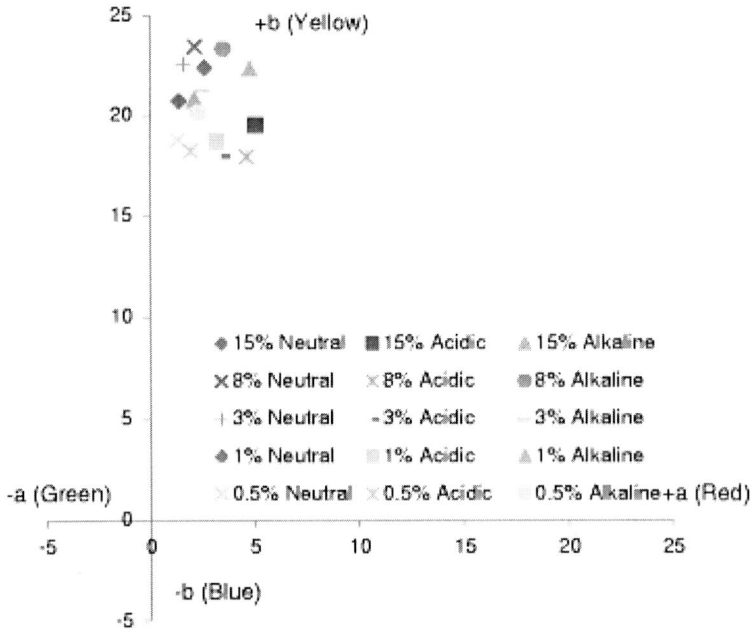

Figure 5. *a**-*b** plot for alum-mordanted dyed woolen fiber

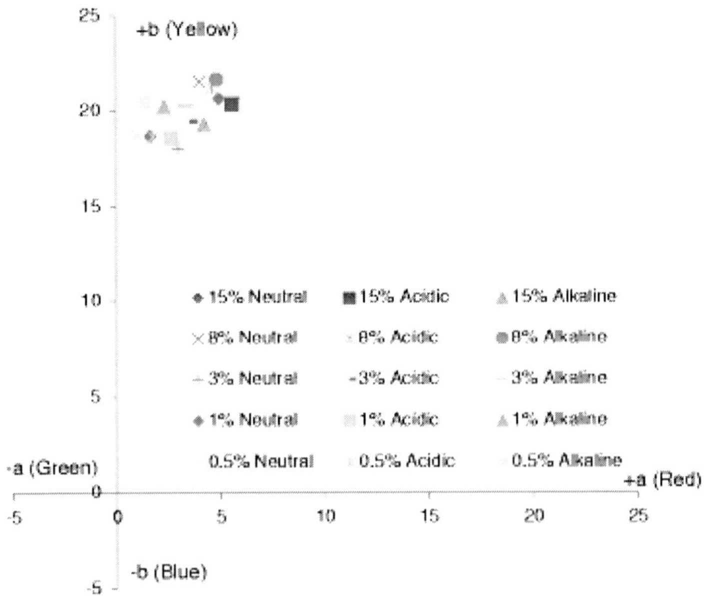

Figure 6. *a**-*b** plot for tin-mordanted dyed woolen fiber

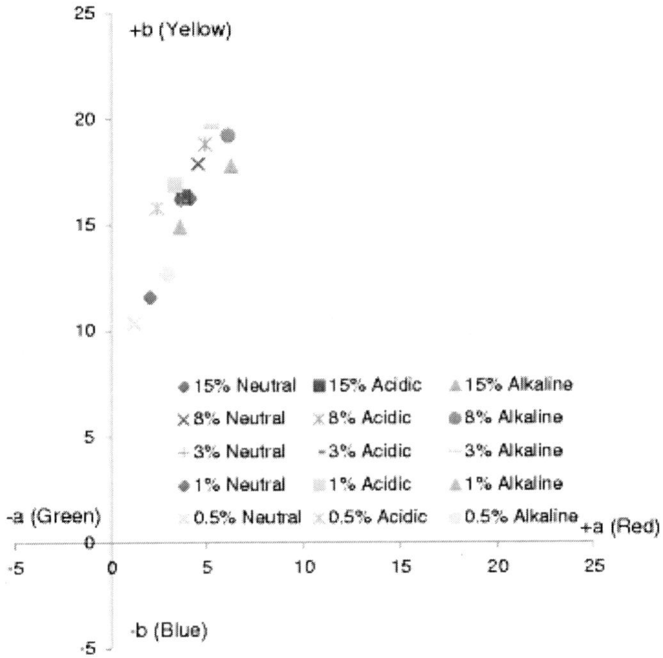

Figure 7. *a**-*b** plot for iron-mordanted dyed woolen fiber

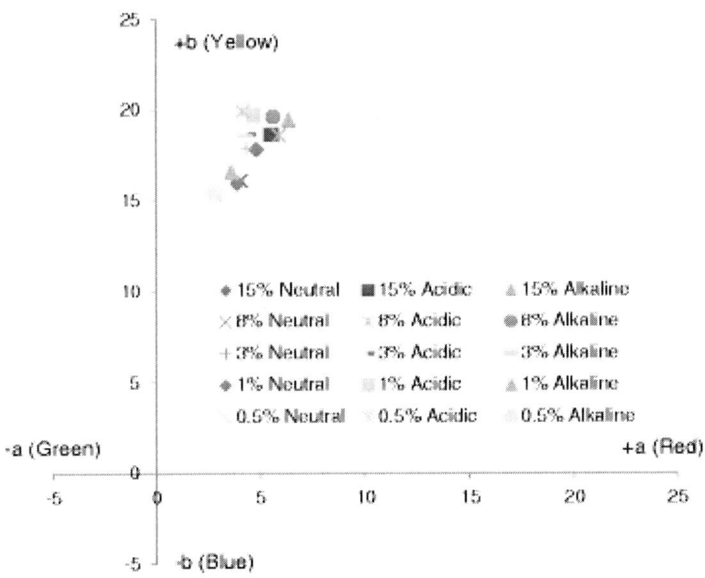

Figure 8. *a**-*b** plot for control-dyed woolen fiber

Table 5. Shade cards for *Terminalia chebula* natural dye

Dye conc.	Dyeing and treatment medium pH	Un-mordanted	Mordant used		
		Control	Alum	Iron	Tin
15%	(Neutral pH-7)				
	(Acidic pH-4)				
	(Alkaline pH-9)				
8%	(Neutral pH-7)				
	(Acidic pH-4)				
	(Alkaline pH-9)				
3%	(Neutral pH-7)				
	(Acidic pH-4)				
	(Alkaline pH-9)				
1%	(Neutral pH-7)				
	(Acidic pH-4)				
	(Alkaline pH-9)				
0.5%	(Neutral pH-7)				
	(Acidic pH-4)				
	(Alkaline pH-9)				

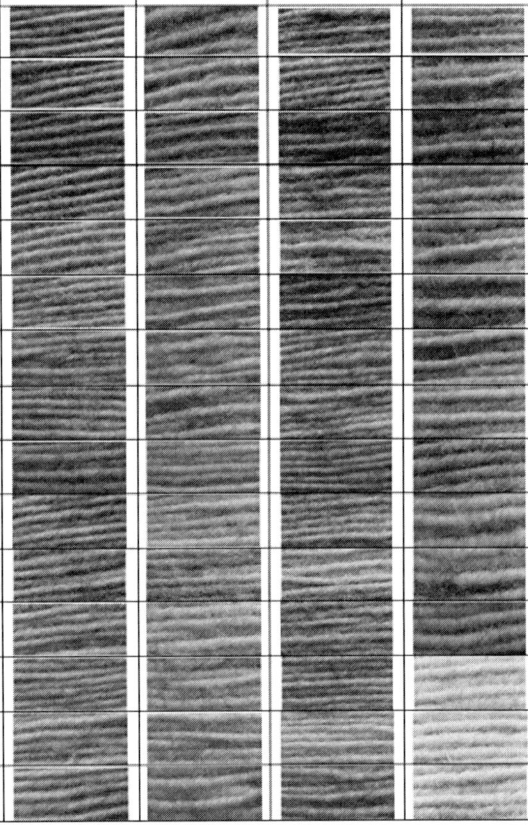

3.9 Wash fastness

From the wash fastness data in Tables 1, 2, 3, and 4, all the *T. chebula*-dyed woolen yarns (un-mordanted as well as mordanted) showed a good to excellent wash fastness rating of 3–5, and negligible staining on adjacent fabrics (cotton and wool) was observed. Relatively good fastness to washing for un-mordanted dyed woolen yarn is attributed to the affinity of coloring components to the yarn in the form of H bonding, ionic bonding, and van der Waals forces (Khan et al., 2015; Yusuf et al., 2015).

3.10 Rub fastness

Un-mordanted and mordanted dyed woolen yarns were tested for dry and wet rub fastness properties and were found more or less the same (4–5 on the grayscale) for both un-mordanted and mordanted dyed samples. Rub fastness of alum-mordanted samples were relatively found somewhat less than that of un-mordanted and mordanted with iron or tin which was due to weak coordination complexation of aluminum ions.

4. CONCLUSIONS

This study evaluated the color parameters (*K/S*, *L**, *a**, *b**) and fastness properties of woolen yarns dyed with*T. chebula* natural dye. The use of different metal salt mordants produced a wide range of beautiful shades with good color and wash fastness (light, wash, and rub fastness) properties. It was found that the *K/S*values of dyed woolen yarn were found higher in the case of iron mordants and lightness was increased with alum and tin mordants. Furthermore, it was found that alum mordants affected negatively the light fastness. In the end, it can be said that *T. chebula* natural dye can provide bright hues with good color fastness properties with different types of mordants.

ACKNOWLEDGEMENTS

Financial support provided by the University Grants Commission, Govt. of India, through the Non-NET Fellowship for Ph.D. students (MS and MNB) and the BSR Research Fellowship in Sciences for Meritorious students (SI and LJR), is highly acknowledged.

COMPETING INTERESTS

The authors declare that they have no competing interests.

AUTHORS' CONTRIBUTIONS

MS and FM designed the experiment. MS conducted most of the experiments. SI interpreted the experimental data and drafted the manuscript. MNB helped MS in performing experiments on dyeing studies of the Harda plant. LJ helped to analyze the experiment. MAK provided technical help. All authors read and approved the final manuscript.

REFERENCES

1. Dalby, G. (1993). Greener mordants for natural coloration. *Journal of the Society of Dyers and Colourists, 109*(1), 8–9.
2. Islam, S., & Mohammad, F. (2015). Natural colorants in the presence of anchors so-called mordants as promising coloring and antimicrobial agents for textile materials. *ACS Sustainable Chemistry & Engineering,3*(10), 2361–2375.
3. Jothi, D. (2008). Extraction of natural dyes from African marigold flower (Tagetes erecta L) for textile coloration.*Autex Research Journal, 8*(2), 49–53.
4. Khan, M. A., Khan, M., Srivastava, P. K., & Mohammad, F. (2005). Extraction of natural dyes from myrobalan, gallnut and pomegranate, and their application on wool. *Colourage, 52*(12), 53–60.

5. Khan, M. I., Khan, S. A., Yusuf, M., Shahid, M., Mohammad, F., & Khan, M. A. (2010). Eco-friendly shades on wool using mixed mordants with Acacia catechu (Cutch). *Colourage, 57*(8), 81–88.

6. Khan, M. I., Ahmad, A., Khan, S. A., Yusuf, M., Shahid, M., Manzoor, N., et al. (2011). Assessment of antimicrobial activity of catechu and its dyed substrate. *Journal of Cleaner Production, 19*(12), 1385–1394.

7. Khan, S. A., Ahmad, A., Khan, M. I., Yusuf, M., Shahid, M., Manzoor, N., et al. (2012). Antimicrobial activity of wool yarn dyed with Rheum emodi L. (Indian Rhubarb). *Dyes and Pigments, 95*(2), 206–214.

8. Khan, S. A., Islam, S., Shahid, M., Khan, M. I., Yusuf, M., Rather, L. J., et al. (2015). Mixed metal mordant dyeing of wool using root extract of *Rheum emodi* (Indian Rhubarb/Dolu). *Journal of Natural Fibers, 12*(3), 243–255.

9. Maulik, S. R., & Agarwal, K. (2014). Painting on handloom cotton fabric with colourants extracted from natural sources. *Indian Journal of Traditional Knowledge, 13*(3), 589–595.

10. Mihalick, J. E., & Donnelly, K. M. (2006). Using metals to change the colors of natural dyes. *Journal of Chemical Education, 83*(10), 1550–1551.

11. Naik, G. H., Priyadarsini, K. I., Naik, D. B., Gangabhagirathi, R., & Mohan, H. (2004). Studies on the aqueous extract of Terminalia chebula as a potent antioxidant and a probable radioprotector. *Phytomedicine, 11*(6), 530–538.

12. Pfundstein, B., El Desouky, S. K., Hull, W. E., Haubner, R., Erben, G., & Owen, R. W. (2010). Polyphenolic compounds in the fruits of Egyptian medicinal plants (Terminalia bellerica, Terminalia chebula and Terminalia horrida): characterization, quantitation

and determination of antioxidant capacities. *Phytochemistry, 71*(10), 1132–1148.

13. Prabhavathi, R., Devi, A. S., & Anitha, D. (2014). Improving the colour fastness of the selected natural dyes on cotton. *IOSR Journal of Polymer and Textile Engineering, 1*(4), 21–26
14. Samantha, A.K., & Agarwal, P. (2009). Application of natural dyes on textiles, Indian Journal of Fibre & Textile Research, 34(December), 384–399.
15. Shahid, M., Ahmad, A., Yusuf, M., Khan, M. I., Khan, S. A., Manzoor, N., et al. (2012). Dyeing, fastness and antimicrobial properties of woolen yarns dyed with gallnut (Quercus infectoria Oliv.) extract. *Dyes and Pigments, 95*(1), 53–61.
16. Shahid, M., Islam, S., & Mohammad, F. (2013). Recent advancements in natural dye applications: a review.*Journal of Cleaner Production, 53*(August), 310–331.
17. Uddin, M. G. (2014). Effects of different mordants on silk fabric dyed with onion outer skin extracts.*Journal of Textiles, 2014*, 1–8.
18. Vankar, P. S., Shanker, R., Mahanta, D., & Tiwari, S. C. (2008). Ecofriendly sonicator dyeing of cotton with Rubia cordifolia Linn. using biomordant. *Dyes and Pigments, 76*(1), 207–212.
19. Yusuf, M., Shahid, M., Khan, S. A., Khan, M. I., Islam, S.-U., Mohammad, F., et al. (2013). Eco-dyeing of wool using aqueous extract of the roots of Indian Madder (*Rubia cordifolia)* as natural dye.*Journal of Natural Fibers, 10*(1), 14–28.
20. Yusuf, M., Shahid, M., Khan, M. I., Khan, S. A., Khan, M. A., & Mohammad, F. (2015). Dyeing studies with henna and madder: a research on effect of tin (II) chloride mordant. *Journal of Saudi Chemical Society, 19*(1), 64–72.

21. Zhou, Y., Zhang, J., Tang, R., & Zhang, J. (2015). Simultaneous dyeing and functionalization of silk with three natural yellow dyes. *Industrial Crops and Products, 64*(February), 224–232.

REMOVAL OF HEAVY METAL CHROMIUM FROM TANNERY EFFLUENT USING ULTRAFILTRATION MEMBRANE

P. Angelin Vinodhini and P. N. Sudha

P.G. and Research Department of Chemistry, D.K.M College for Women, Vellore, Tamil Nadu, India

ABSTRACT

The present study was aimed to fabricate a novel ultrafiltration membrane using cellulose acetate, nanochitosan, and polyethylene glycol of ratio 1:2:2 by phase inversion method. Analytical techniques such as Fourier transform infrared spectroscopy (FTIR), X-ray diffractometer (XRD), and scanning electron microscopy (SEM) have been employed to characterize the prepared membrane. FTIR and XRD results revealed the formation of blended membrane with increased amorphous nature. SEM studies exhibited the rough surface with numerous pores. The prepared membrane was also characterized for its ultrafiltration performance by membrane compaction, pure water flux, water content, and porosity. The results showed that the membrane obtained a pure water flux of 25.32 l/m^2 h. The water content was found to be 24 % and a high porosity of 83 % was obtained which showed the membrane's high hydrophilicity and more porous nature. The main focus of the present study was toxic hexavalent chromium removal from tannery effluent using the prepared hydrophilic membrane, which causes various adverse effects. The effect of pH of the solution (viz., pH

5, 7, and 9), membrane thickness (0.1 and 0.2 mm), and the applied pressure (50 and 100 kPa) were studied which are the key factors in determining the efficiency of the prepared membrane in remediation of tannery effluent. The results showed high percentage removal of chromium at pH 7 using 0.2 mm thickness at 100 kPa. The physicochemical parameters of the tannery effluent were also found to be reduced.

KEYWORDS

Cellulose acetate, Nanochitosan, Polyethylene glycol, Ultrafiltration membrane, Remediation of tannery effluent

1. BACKGROUND

The one and only substance that is found on earth's crust for the survival of living things is water which regulates the earth's temperature as well as the temperature of the human body. The most valued water is now get polluted by the anthropogenic activities by the emission of domestic or urban sewage, agricultural waste, pollutants, and industrial effluents. There are numerous data and statistics which claim that the earth's water resources are being depleted, polluted, and rendered unpotable at an alarming rate (Malik et al.2014). Thus, the whole world is under stress for the want of pure water. The main source of water pollution is industrial effluents which consist of acids, alkalis, toxic heavy metals, oil, grease, dyes, pesticides, and even radioactive materials.

Heavy metals which are the natural constituents of the earth crust are persistent environmental contaminants because they cannot be degraded or destroyed (Duruibe et al. 2007). Heavy metals such as mercury, lead, cadmium, copper, and chromium are toxic even in small amounts (Pang et al. 2015). The health hazards associated with exposure to chromium are dependent on its oxidation state. The hexavalent form, Cr(VI), is toxic and has adverse effects on the skin like ulcerations, dermatitis, and allergic skin reactions. It is also proved to be carcinogenic (Barbooti 2015). The chromium removal techniques available are precipitation, adsorption, and solvent extraction (Moosa et al. 2015) among which membrane techniques have been proven to be one of the respective methods. It is simple and economical for the removal of heavy metal ions from aqueous solution.

A wide range of polymeric materials have been used to prepare various membrane types. The choices of polymer taken for the current study are the biopolymer chitosan in its nanosize, a porogen polyethylene glycol, and membrane matrix material cellulose acetate. Each polymer has their unique properties constituting for the performance of the membrane prepared. The second most abundantly available chitin is deacetylated to get chitosan (Sudha 2010) which is further converted into its nanoform by suitable crosslinking agents. Owing to its invaluable applications, chitosan (CS) has become very vital in various fields like pharmaceutical (Kato et al. 2003; Kumar et al. 2004), biomedical (Suh and Matthew 2000; Tucci and Rocotti 2003; Ng and Swami 2005), water treatment (Northcott et al. 2005; Crini 2005), cosmetics (Rinoudo 2006; Sun et al. 2006), agriculture (Boonlertniruni et al. 2008; El Hadrami et al. 2010), and food industry (Ham-Pichavant et al. 2005; De Lima et al. 2010). The properties of chitosan are found to be more enhanced when used in its nanoscale. Thus, the ever increasing field of nanotechnology has also been implemented in the present study.

Cellulose acetate (CA), the most important cellulose ester was one of the first membrane polymer used for aqueous-based separations and used both as reverse osmosis and UF membrane material (Kuttowy and Sourirajan 1975). Its convincing properties such as good toughness, high biocompatibility, good desalting, high flux, and relatively low cost (Idris and Yet 2006) makes it versatile for use as membrane matrix. Another important additive for membrane is the pore forming agent which is very essential for the formation of pores in the membrane. The synthetic polyether polyethylene glycol (PEG) was employed as porogen which is inexpensive, non-toxic, and also soluble in many organic solvents like toluene, ethanol, acetone, and chlorinated hydrocarbons. However, it is insoluble in ethers, aliphatic hydrocarbons, fats, and mineral oil (Rowe et al. 2003). So far, there is no literature found for using nanochitosan in membrane technology which emphasizes the novelty of the present work.

Thus, the main objective of the present work is to prepare a novel ultrafiltration membrane of nanochitosan, cellulose acetate, and polyethylene glycol using popular phase inversion process, to characterize for its formation and suitability using Fourier transform infrared spectroscopy (FTIR), X-ray diffractometer (XRD), scanning electron microscopy (SEM), membrane compaction, pure water flux, water content, and porosity and to find the efficiency of the prepared membrane for the removal of chromium from tannery effluent and the analysis of the results to achieve the most favorable conditions.

2. METHODS

2.1 Materials

Commercial grade cellulose acetate and sodium lauryl sulfate (SLS) were obtained from ChenChems, India. Chitosan was purchased from India Sea Foods, Cochin, Kerala, which is 92 % deacetylated. Polyethylene glycol and acetic acid were procured from Sisco Research Laboratories Pvt. Ltd., Mumbai, India. *N,N′*-dimethylformamide was procured from Thermo Fisher Scientific India Pvt. Ltd., Mumbai. Sodium tripolyphosphate (TPP) was procured from Finar Chemicals Limited, Ahmedabad, India. Deionized and distilled water was employed for the preparation of nanochitosan and gelation bath.

2.2 Preparation of nanochitosan

Nanochitosan was prepared by sol-gel method using TPP. Ten grams of chitosan was dissolved in 2000 ml of 2 % (*v*/*v*) acetic acid. The above solution was stirred well for 45 min. Then, 100 ml of 0.75 mg/ml TPP was dropped into the beaker. A milky emulsion was obtained. The nanochitosan was washed several times in deionized water. This solution was then allowed to settle for 24 h. The supernatant solution was then decanted (Sivakami et al. 2013). Particle size of the prepared nanochitosan was investigated by using a dynamic light scattering (DLS) device (zeta plus from Brookhaven), and the average particle size of nanochitosan was found to be 100 nm.

2.3 Fabrication of membranes

2.3.1 Solution blending of polymers

Membrane was prepared using CA, NCS, and PEG in 1:2:2 ratio with a polar solvent dimethylformamide (DMF) under constant mechanical stirring for 1 h at room temperature. The homogeneous polymer solution obtained was allowed to stand for 10 min in an airtight condition to get rid of air bubbles.

2.3.2 Casting of membranes

The membrane was prepared by phase inversion process (Witte et al. 1996; Loeb and Sourirajan 1963; Arthanareeswaran et al. 2007; Nath 2008; Velu et al. 2011) which is a well-known technique to prepare asymmetric membranes. The deaerated polymeric solution was casted on a glass plate, and the thickness of the membrane was maintained about 0.2 mm approximately. After allowing the solvent to evaporate, the glass plate was immersed in the gelation bath for final precipitation.

Prior to membrane casting, 2 l gelation bath consisting of 2.5 % (*v*/*v*) DMF and 0.2 wt.% SLS in distilled water (non-solvent) was prepared and cooled to 15 ± 2 °C (Arthanareeswaran et al. 2004). After gelation process, the membrane was removed from it and washed thoroughly with demineralized water and subsequently stored in 0.1 % formalin solution to prevent microbial growth.

2.4 Membrane characterization

FTIR

A thermo nicolet AVATAR 330 IR spectrophotometer was used to record the IR spectra within the range of 4000–400 cm^{-1}. The IR spectra were recorded in a solid state using a KBr pellet method.

XRD

The molecular packing of the blend membranes were analyzed with X-ray powder diffractometer (XRD-SHIMADZU XD-D1) using a Ni-filtered Cu Kα X-ray radiation source. Percentage of crystallinity X_c% was measured as a ratio of crystalline area to the total area of the peaks using the formula given by Nara and Komiya (1983).

SEM

The surface and cross-sectional morphology of the prepared blend membranes were viewed and photographed using scanning electron microscopy. For the analysis, the blend membranes were cut into pieces of various sizes and wiped with filter paper. The membranes were coated with a thin gold-palladium layer by a sputter coater unit (VG-Microtech, UCK field, UK), and the cross section topography was analyzed with a Cambridge stereoscan 440 scanning electron microscope (SEM, Leica, Cambridge, UK) operated at an acceleration voltage of 20 kV.

2.5 Ultrafiltration characterization

2.5.1 Membrane compaction

The prepared membranes were thoroughly washed and cut into desired size and loaded in the UF test cell of 38.5 cm^2 area and initially pressurized with distilled water at 240 kPa. The water flux was measured at every 10 min. The flux generally declines initially and attained steady state after 90 min of compaction. The prepressurized membranes were used in subsequent ultrafiltration experiments at 100 kPa (Kuttowy and Sourirajan 1975).

2.5.2 Pure water flux

Membranes after compaction were subjected to pure water flux estimation at a transmembrane pressure of 100 kPa. The flux was measured under steady state flow. The pure water flux was determined (Osada and Nakagawa 1992) using the expression,

$$J_w = \frac{Q}{\Delta t \times A}$$

where J_w is the pure water flux (l m^{-2} h^{-1}), Q is the amount of permeates collected (l), Δt is the sampling time (h), and A is the membrane area (m^2).

2.5.3 Water content

Water content of the membranes were obtained after soaking membranes in water for 24 h and the membranes were weighed followed by mopping it with blotting paper. These wet membranes were placed in vacuum drier at 75 °C for 48 h and cooled it to room temperature in a dessicator, and the dry weights of the membranes were determined. The percent water content was calculated (Tamura et al. 1981) by using the following equation:

$$\%\ \text{water content} = \frac{\text{wet sample weight} - \text{dry sample weight}}{\text{wet sample weight}} \times 100$$

2.5.4 Porosity

The overall porosity (ε) was determined by gravimetric method, as defined in the following equation (Li et al.2009):

$$\varepsilon = \frac{w_1 - w_2}{A \times l \times d_w}$$

where w_1 is the weight of the wet membrane, w_2 is the weight of the dry membrane, A is the membrane effective area (m^2), d_w is the water density (0.998 g/cm^3), and l is the membrane thickness (m).

2.6 Collection of tannery effluent from Ranipet, Vellore District, Tamilnadu

The effluent sample for the current study was collected from the final discharge point wherein effluent from all the stages of processing were treated and released together from a leather tanning unit in Ranipet, Vellore district, Tamilnadu, India, to the water stream. Polythene containers of 5 l capacity was used for the collection of effluent samples and immediately transported to the laboratory with due care and stored at 4 °C in refrigerator until further analysis.

The color of the tannery effluent was observed visually and recorded, and the odor was detected by smelling. pH was measured using Elico pH meter on the site. Other physicochemical parameters such as pH, electrical conductivity (EC), dissolved oxygen (DO), biochemical oxygen demand (BOD), chemical oxygen demand (COD), total dissolved solids (TDS), total solids (TS), total hardness (TH), salinity, turbidity, and sodium were analyzed in the laboratory following standard methods (APHA 1993; Trivedy and Goel1986).

Remediation of tannery effluent using prepared blend membrane by studying the effect of pH of the solution, membrane thickness, and applied pressure

The effluent that has already undergone primary and secondary treatment process in industry itself and that which is let out to the water streams was collected and subjected to subsequent ultrafiltration process. The UF cell was fixed with the prepared membrane and was filled with the tannery effluent solution, and filtration was started. The permeate was collected periodically in the stoppered bottles, and the physicochemical factors were analyzed after each period of remediation as per the methods of APHA (1993) and Trivedy and Goel (1986). The heavy metal chromium was analyzed using Atomic absorption spectrometer (Varian 4AA Atomic Absorption Spectrometer).

The percentage chromium rejection was calculated from the concentration of chromium in the feed and permeates using the equation given below.

$$\% \text{ Cr rejection} = \left[1-\left(C_p/C_f\right)\right] \times 100$$

where C_p is the concentration of chromium in permeate and C_f is the concentration of chromium in feed.

The extent of remediation of the effluents was investigated by changing the pH of the solution, membrane thickness, and applied pressure. The physicochemical parameters of the effluent before and after treatment were measured as mean of three individual observations. The physical and chemical characteristics of the raw effluent were analyzed and given in Table 1 along with the accepted limit given by ISI standards for disposal of industrial wastewater of inland surface water.

2.7 Effect of pH

pH is the most important of all the factors affecting the removal of heavy metals. pH adjustment is extremely important as well since it can be used as a means to selectively separate heavy metals from multicomponent mixtures and the removal rate to separate heavy metals differs when pH is carefully adjusted (Zeng et al. 2009). The effect of pH on the filtration of tannery effluent by the membrane was carried out at room temperature and at three different pH values.

Table 1. Physicochemical analysis of tannery effluent

Parameters	Raw effluent	Accepted limits[a]
Physical parameters		
Color	Blackish color	Colorless
Odor	Disagreeable smell	Odorless
pH	7.43	6.5–8.5
Chemical parameters		
Electrical conductivity (EC) (mS/cm)	27.3	0.288
Dissolved oxygen (DO) mg/l	0.2	4–6
Biochemical oxygen demand (BOD) mg/l	2556	30
Chemical oxygen demand (COD) mg/l	6419	250
Total dissolved solids (TDS) mg/l	15,900	2100
Total solids (TS) mg/l	31,740	2200
Total hardness (TH) mg/l	923	300
Salinity (mg/l)	16,200	Not mentioned
Turbidity (NTU)	453	10
Sodium (mg/l)	5000	1000–1500
Chromium (Cr) (mg/l)	1055	2.0

Values expressed as mean of six individual values

[a]ISI standards for disposal of industrial wastewater of inland surface water

2.7.1 Effect of membrane thickness

Thickness is a key parameter in membrane processes since it influences both the membrane selectivity and the membrane resistance against flux (Abdelrasoul et al. 2015). It also plays an important role in the amount of conductive heat loss though the membrane. The filtration study was conducted using two different membrane thicknesses, viz., 0.1 and 0.2 mm.

2.7.2 Effect of applied pressure

The transmembrane pressure is the driving force for the pressure-driven membrane process (Gökşen 2005). Applied pressure is considered as one of the most critical aspects in the membrane performance and operation. The UF experiment was carried out at two different pressures namely 50 and 100 kPa.

3. RESULTS AND DISCUSSION

3.1 Membrane characterization

FTIR

IR spectroscopy is an excellent method and has been extensively used for the qualitative analysis in which specific absorption peaks can be identified for particular groups (Qu et al. 2010). The interactions between the functional groups in a polymer blend can be obtained directly and gross structural details can be investigated. The chemical nature of a substance including chemical bonds, molecular orientations, molecular energy levels, and molecular interactions (Gorey and Escobar 2011) can be assessed using this technique. Thus, the prepared blend has been analyzed using FTIR, and the possible interactions that had taken place were studied.

The FTIR spectral details of the prepared NCS are shown in Fig. 1. A prominent broad peak at 3385.92 cm^{-1}indicates the presence of O–H and N–H stretching vibration (Vimal et al. 2013). Peak at 1324.83 cm^{-1} is assigned to P=O stretching vibration whereas at 1038.87 cm^{-1} is assigned to P–O stretching vibration. The N–H bending vibration corresponds to the peak at 1510.05 cm^{-1} which was proved by Mohammadpour Dounighi et al. (2010) who obtained peak at 1540 cm^{-1} in CS/TPP nanoparticles. The deformation vibrations of C–H and O–H groups were observed at 1431.02 and 1376.80 cm^{-1}. A peak for C=O stretching vibration is observed at 1635.20 cm^{-1}. Peaks at 1219.00,

1163.38, and 1113.24 cm^{-1} were due to C–N stretching, C–O stretching, and C–O–C ether linkages, respectively.

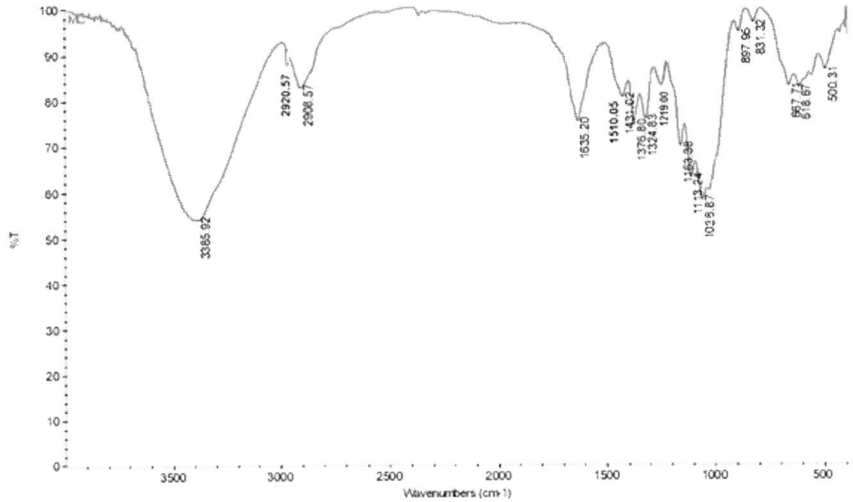

Figure 1. FTIR spectrum of nanochitosan

The FTIR spectral details of CA/NCS/PEG (1:2:2) blend membrane are given in Fig. 2. The O–H and N–H stretching vibration of the membrane is observed at 3389.80 cm^{-1} which is wider. The O–H stretching is attributed to the enhanced intermolecular hydrogen bonding formed between the polymers (Yu et al. 1999). The peaks at 2948.54 cm^{-1} and 1642.36 cm^{-1} were due to the asymmetric C–H stretching and C=O stretching vibration, respectively. The peak at 1495.20 cm^{-1} was assigned to the deformation vibration of N–H. A peak for P=O stretching vibration was found at 1284.14 cm^{-1} and for P–O stretching vibration at 1014.99 cm^{-1} which showed a shift from 1038.87 cm^{-1} characteristic of pure NCS (Vijayalakshmi et al.2014). C–N stretching vibration and C–O–C ether linkage were observed at 1220.51 and 1056.94 cm^{-1}, respectively.

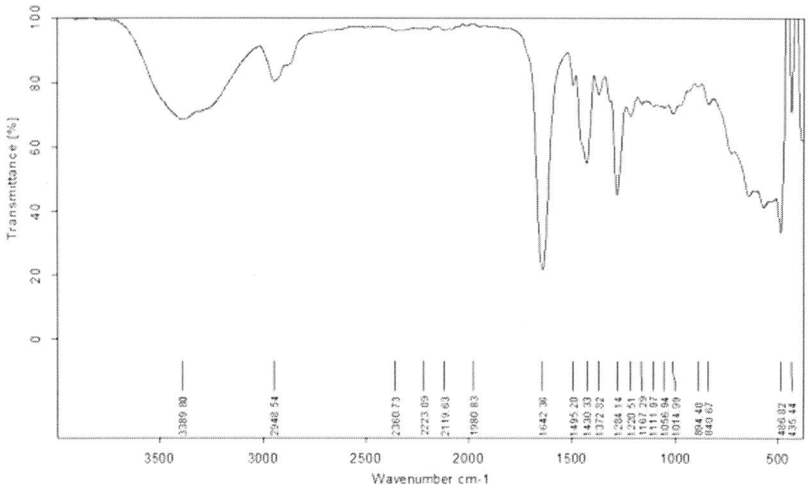

Figure 2. FTIR spectrum of CA/NCS/PEG (1:2:2) blend membrane

There is a shift in the OH and NH frequencies for the prepared membrane from NCS, i.e., the peak obtained at 3385.92 cm^{-1} has been shifted to 3389.80 cm^{-1}. The C=O peak at 1635.20 cm^{-1} for NCS has been found to be shifted to 1642.36 cm^{-1}. Peak at 1324.83 cm^{-1} for P=O was observed to have a shift in the IR spectrum to 1284.14 cm^{-1}. This shows that there is strong intermolecular hydrogen bonding taking place which leads to the membrane formation. Moreover, the P–O bond at 1038.87 cm^{-1} for NCS has undergone a significant shift in the prepared membrane and observed at lower region of 1014.99 cm^{-1}. Thus, the FTIR spectra analysis reflects that the nanoparticles were binded with the CA matrix due to strong interactive forces.

XRD

The X-ray diffraction analysis is a non-destructive analytical technique that gives the information about the crystallographic structure, chemical composition, and physical properties of materials (Subbu et al. 2014). It has been routinely used for crystalline phase identification based on the diffraction peak position and pattern (Jayakrishnan and Ramesan 2016). The crystalline/amorphous ratio of the constituent polymer in a blend plays a vital role in understanding the development of membranes during gelation. Thus, X-ray diffraction patterns of the prepared membrane were measured to investigate the change in the crystalline nature.

Figure 3 shows the X-ray diffractogram of NCS. The percentage of crystallinity was found to be 17 % which may be due to the certain orderliness in its structure which was retained even after the addition of the crosslinker TPP to chitosan. The 2θ value obtained at 26° was wider indicating the amorphous nature of the chitosan nanoparticles of size 100 nm. For the chitosan nanoparticles which were in the range of 5–180 nm, the 2θ value was found to be 20.4357 as reported by Vellingiri et al. (2013).

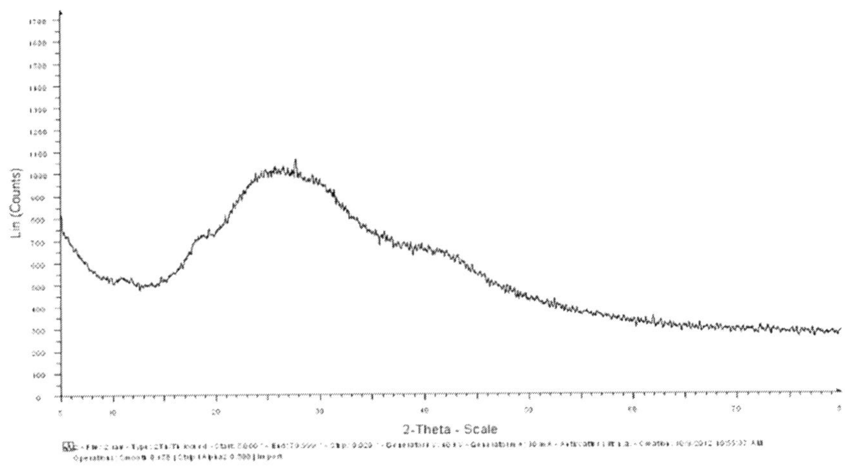

Figure 3.X-ray diffractogram of NCS.

Figure 4 shows the X-ray diffractogram of CA/NCS/PEG (1:2:2) blend membrane. The blend showed lesser percentage of crystallinity which is 6 % when compared to NCS whose value was 17 %. This revealed that the blend has become more amorphous which may be due to enhanced disruption of orderliness by the addition of nanochitosan and polyethylene glycol in greater amount. It was observed that nanochitosan was amorphous and showed diffraction peaks at 2θ

values 15.19° and 23° (Karuppasamy et al. 2012). There was also a significant shift in the 2θ values to 10.5° and 29.5° which were fairly broad. It was also observed that the intensity of the crystalline peaks was decreased and broadened upon the addition of nanochitosan. This disorderliness may be attributed to the formation of intermolecular hydrogen bonding formed between CA, NCS, and PEG, thus favoring the filtration process.

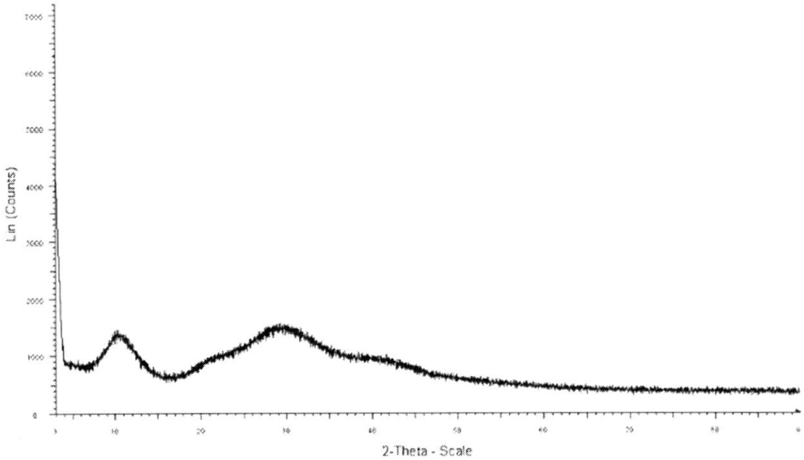

Figure 4.X-ray diffractogram of CA/NCS/PEG (1:2:2) blend membrane

3.2 Scanning electron microscopy

The scanning electron microscope is an extremely useful tool to characterize the surface morphology of the membrane (Zeng et al. 2016) and to investigate the structure of polymeric integrally skinned membrane (Kesting 1985).

The surface morphology of nanochitosan and the blend membrane were observed using SEM and are shown in Figs. 5 and 6. The surface of the membrane helps to identify the significance of membrane in the mechanism of selectivity and permeability. The top layer of membrane is responsible for the permeation whereas the sublayer acts as a mechanical support (Rajesh et al. 2011). The surface of the prepared membrane clearly shows their asymmetric nature and roughness. The membrane has uniform pores exhibiting more porous structure.

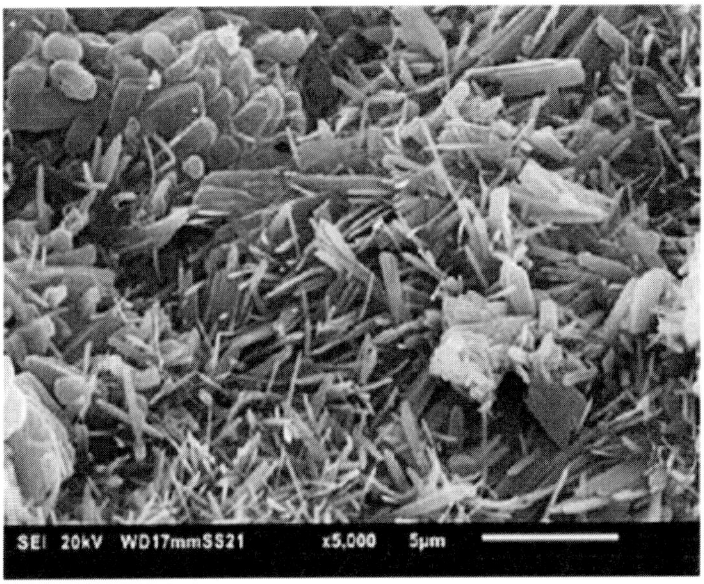

Figure 5. SEM micrograph of nanochitosan

The shape and structure were related to the concentration and composition of additive in the casting solution (Sivakumar et al. 2000). Thus, the prepared membrane was found to be more porous with more PEG content showing small multiple uniform voids when compared with NCS. This is supported by the study done by Saljoughi et al. (2009) that increasing PEG concentration causes formation of greater macrovoids and more porous structures due to the intensification of thermodynamic instability of the casting solution by PEG. Moreover, the number of pores and pore sizes depends upon the pore forming agent PEG, which leaches out of the membrane in gelation bath during phase separation process (Ananth et al. 2012). However, the pore size of membrane was reduced with increased PEG content in the blend solution (Khulbe et al. 1995) due to the reduced leachability of PEG during gelation in view of strong interactions of PEG with CA and NCS. Hence, the membrane can facilitate the filtration process efficiently.

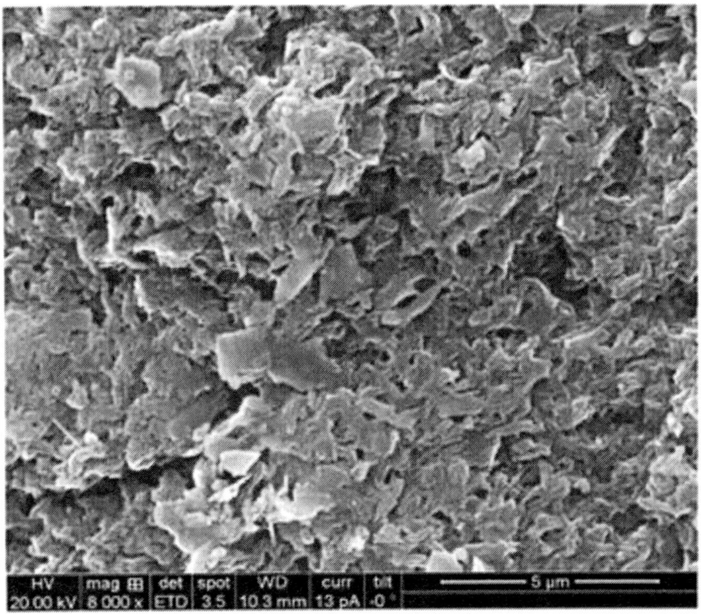

Figure 6. SEM micrographs of CA/NCS/PEG (1:2:2) blend membrane

3.3 Ultrafiltration characterization of the membrane

The efficiency of the prepared membrane was found out using various ultrafiltration characterization techniques such as membrane compaction, pure water flux, water content and porosity.

3.3.1 Membrane compaction

Compaction occurs when membranes are subjected to high pressures. The significance of compaction varies with membrane formulation and operating pressure (Pusch and Mossa 1978). Before using a fresh membrane, it was compacted to make membrane with rigid pore structure and size. When compacted with deionized water for 2 h at a transmembrane pressure of 240 kPa which is higher than the maximum operating pressure used for the present study, there may be reorganization of the polymeric chains (Maheswari et al. 2013). The limitation of the transmembrane pressure was <300 kPa.

The water flux was calculated from the experimental permeate flow rate measured at every 10 min interval. During compaction, initially, the pure water flux was found to be high and declines gradually and reaches a steady state after 90 min. This constant value indicates the completion of compaction. This may be due to the fact that the walls of the pores become closer, denser, and uniform resulting in reduction in pore size as well as the flux during compaction (Eren et al. 2015). Furthermore, the polymeric chains reorganize under hydraulic pressure during compaction leading to change in the structure with a lowered volume of porosity.

3.3.2 Pure water flux

The key requirement of any membrane is pure water flux (PWF), and it was determined for the prepared blend membrane at 100 kPa pressure after completion of compaction. Pure water flux is one of the key components that explain the pore size and pore distribution of the membranes (Shi et al. 2012). It is an essential parameter for the membranes, which is useful for any industrial process. The value of pure water flux of the prepared membrane was found to be 25.32 l/m^2 h.

3.3.3 Water content

The percentage water content represents the fraction of water molecules occupied in the pores of the membrane. This is considered to be an important parameter for membrane characterization (Tamura et al.1981), since PWF can be predicted based on these results also. The percent water content was calculated for the prepared membrane, and the value obtained was 24 %. This shows that the prepared membrane has become more porous due to the leachability of water-soluble additive, PEG, leading to the formation of more number of pores. This shows the membrane's hydrophilicity since water content of the membrane is related to hydrophilicity of the membrane (Sivakumar et al. 2006).

3.3.4 Porosity

Porosity, ε, is the void fraction of the membrane. The membrane porosity is related to the pore formation during the process of membrane formation by immersion precipitation (Wienk et al. 1996). A highest percentage of porosity of 83 % was obtained indicating the more porous nature, and it is attributed to the leachability of PEG present in higher concentration.

3.4 Remediation of tannery effluent using prepared blend membrane by studying the effect of pH of the solution, membrane thickness, and applied pressure

3.4.1 Effect of pH

The effect of pH on the filtration of tannery effluent by the membrane was carried out at room temperature and at three different pH values, and the results obtained are shown in Figs. 7, 8, and 9.

The pH of the tannery effluent was changed to acidic, and the filtration process was carried out. It was found that the physicochemical parameters were greatly reduced after 5 min. With the increase of time, these parameters were found to be increased which may be due to clogging of pores. The percentage removal of chromium at pH 5 was 90.142 % at 5 min and 39.242 % at 95 min.

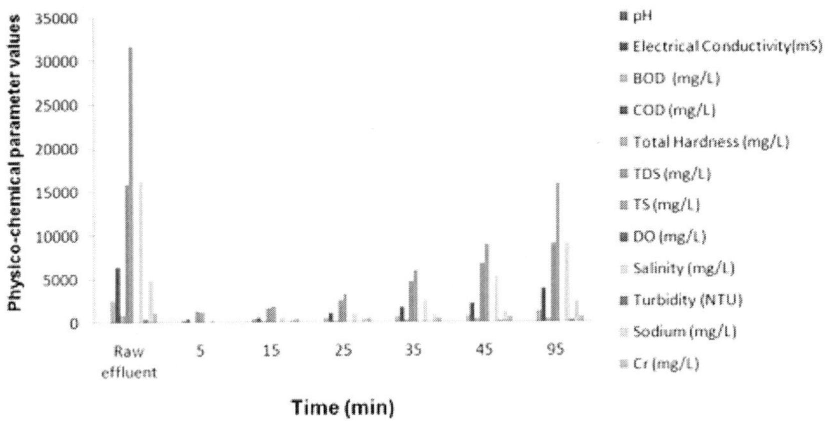

Figure 7. Physicochemical parameters of the treated effluent at pH = 5

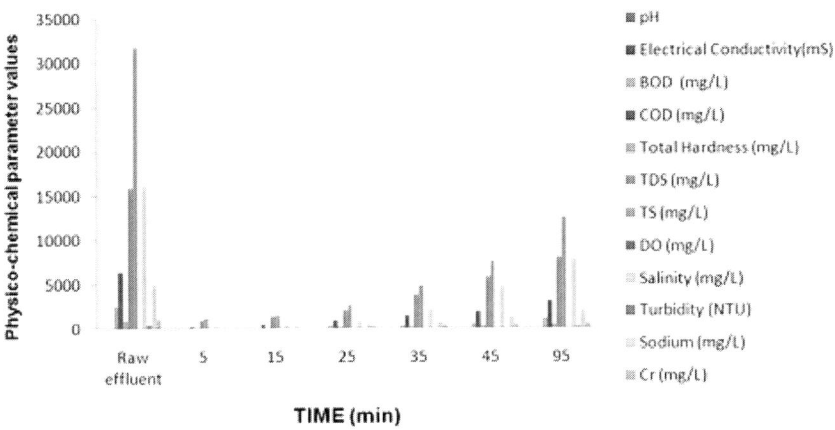

Figure 8. Physicochemical parameters of the treated effluent at pH = 7

The raw effluent itself in the pH of 7 and the maximum removal was found to take place at this pH. The high concentration of Cr discharged by leather industries was found to be 1055 mg/l. This was observed to be reduced to 46 mg/l after 5 min of filtration and to 586 mg/l at 95 min. Also, the other physicochemical parameters have found to be significantly reduced. Some even fell within the permissible limit of inland surface waters as given by ISI. The BOD and COD of the effluent were 2556 ml/l and 6419 mg/l, respectively, which were very high compared to the treated effluent. It was revealed from the results that even after 95 min, the membrane was able to reduce 56.22 % of BOD and 50.97 % of COD. The COD values were found to be hardly decreased as reported by Krishanamoorthi et al. (2009) which may be explained, as COD was mainly soluble consisting in hydrolyzed proteins of low molecular weights, but the effluent has other ionic materials like sodium and chromium.

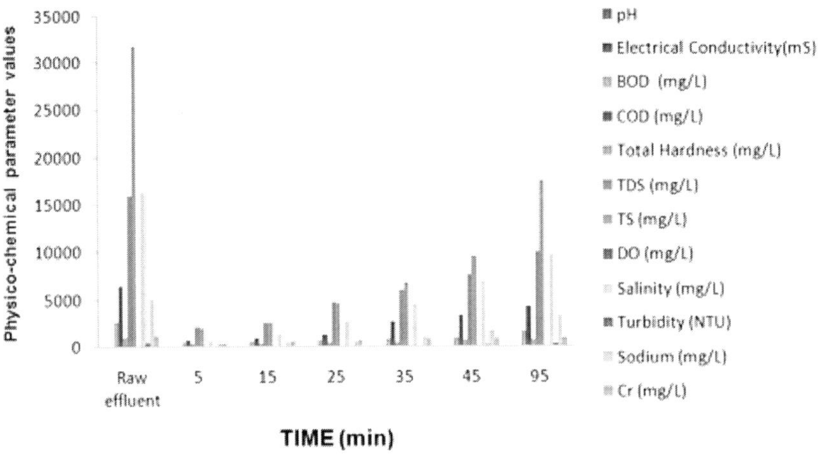

Figure 9. Physicochemical parameters of the treated effluent at pH = 9

In the basic medium, the percentage removal of chromium as well as the physicochemical parameters was not as high when compared with that of acidic and neutral medium. The percentage removal of chromium at pH 9 was only 77.915 % at 5 min which is very less when compared at pH 5 and 7, which are 90.142 and 95.640 %, respectively. This may be because the divalent chromium species $CrO_4{}^{2-}$ which is dominant at high pH would have formed metal complex with sodium present in the effluent and passed through the membrane (Chen et al. 2008).

Figure 10 indicates the percentage removal of chromium in tannery effluent at pH 5, pH 7, and pH 9 during various time intervals. The decrease in chromium rejection percentage with time for the studied pH at 5, 7, and 9 may be presumably due to the saturation of the membranes active sites with chromium ions (Habibi et al. 2015). On comparing the values obtained, it was clear that the removal of Cr was the highest in neutral medium than the acidic medium and was the least at basic medium. As stated by Arthanareeswaran et al. (2007), the rejection increases as the pH of the chromium solution increases and chosen an optimum pH of 6 for further rejection studies. The trend of results obtained for tannery effluent was found similar to that of the removal of Cr using synthetic wastewater which may be attributed to the formation of soluble hydroxides at high pH. Thus, it can be concluded that the membrane can withstand pH variations from acidic to basic. The maximum efficiencies for different parameters were found at neutral pH area, and they gradually reduced on both sides of the neutral pH (Chowdhury et al. 2013). Similar observation was made in the current study which may be accounted that the colloid substances in the leather industrial effluents usually carry negative electrical charges, whereas the membrane possesses a lot of positive charge due to the incorporation of chitosan nanoparticles in the polymer membrane matrix which has the capability of adsorbing particles leading to destabilization and thus also led to the assumption that the membrane acts as a adsorptive membrane.

The effect of pH on chromium rejection plays an important role because the charges of both salt ions and membrane vary with pH (Verliefde et al. 2008). In a study done by Muthukrishnan and Guha (2008), it was found out that the rejection of chromium greatly increased when pH rose from 2 to 7, as more bivalent $CrO_4^{2-}/Cr_2O_7^{\ 2-}$ ions were generated and the polyamide NF membrane was more negatively charged at higher pH. Similar findings were given by Yao et al. (2015) in which a quaternary ammonium-based UF membrane (QA membrane) was used to remove Cr(VI) ions in both acid and neutral pH more efficiently. Hence, removal of chromium from the membrane separation process is reliant on the pH of the solution.

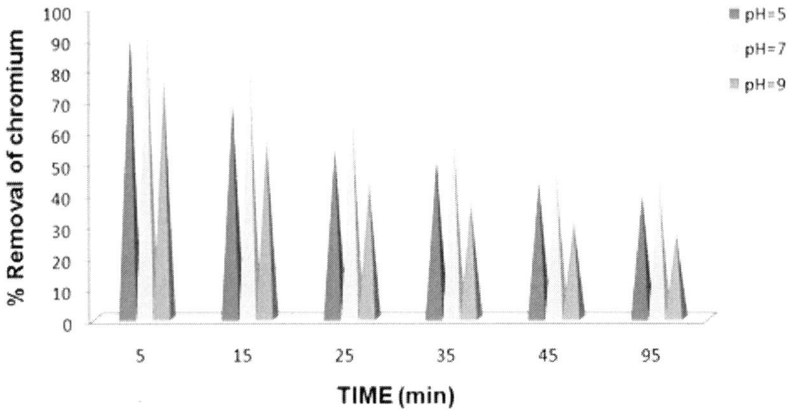

Figure 10. Effect of time on percentage removal of chromium in tannery effluent at various pH

3.4.2 Effect of membrane thickness

The membrane thickness places a crucial role in the removal process. Figure 11 shows the change in various physicochemical parameters at membrane thickness of 0.1 mm.

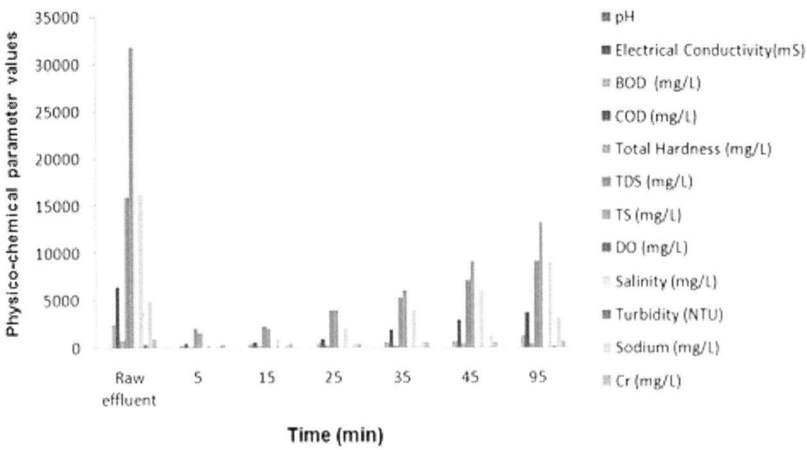

Figure 11. Physicochemical parameters of the treated effluent at thickness = 0.1 mm

The EC, BOD, COD, TH, TDS, TS, DO, salinity, turbidity, Na, and Cr of the raw effluent were drastically high which were found to be reduced to a considerable extent. The EC value of 27.3 mS was observed to be reduced to 7.21 mS during the first 5 min of filtration which was almost 73 % of reduction. Similarly, the values of BOD, COD, TH, TDS, TS, salinity, turbidity, and Na which were 2556, 6419, 923, 15900, 31740, 16200, 453 and 5000 mg/l has been reduced to 312, 513, 223, 2095, 1654, 415, 42 and 218 mg/l, respectively. The percentage of chromium removal was 56.967 % at 5 min and even at 95 min; the membrane has the efficiency to remove about 30.237 % of chromium.

Figure 12 shows the change in the physicochemical parameters when used 0.2 mm of the thickness of membrane. It was evident that there was a vital increase in the removal efficiency of the membrane when its thickness was increased from 0.1 to 0.2 mm. The values of BOD, COD, TH, TDS, TS, salinity, turbidity, and Na which were 312, 513, 223, 2095, 1654, 415, 42 and 218 mg/l, respectively, when used 0.1 mm of membrane were found to be still reduced to 113, 237, 93, 944, 1162, 210, 22 and 112 mg/l, respectively. This enhancement in the removal efficiency may be attributed to thin skin layer and porous sublayer of the asymmetric membrane containing more number of voids. The EC was also retarded from 7.21 to 3.22 mS. In addition, the DO was found to be increased from 0.2 to 4.2 mg/l which favors the life of the aquatic ecosystem since the accepted limit given by ISI was 4–6 mg/l.

Figure 13 shows the effect of time on the percentage removal of chromium using 0.1 and 0.2 mm thickness of the membrane. It can be seen that the percentage removal of chromium with 0.1 mm thickness was 56.967 % which was less when compared with 0.2 mm thickness which gives removal percentage of 95.640. With the increase of time, the removal efficiency was retarded due to pore blocking. This was evidenced by a study conducted by Jyothi et al. (2014) in which the percentage rejection of chromium decreases as time increases. The same trend was observed till the 95 min, i.e., 30.237 % for 0.1 mm and 44.455 % for 0.2 mm thickness. It was obvious that as the membrane thickness increased, the percentage removal of chromium was also increased. An increase in membrane thickness correlates to a decrease in permeation flux and better retention of solute. This could be attributed that membranes had larger pores with increasing thickness

(Abdelrasoul et al. 2015). Due to practical difficulties, using of 0.3 mm thickness has been avoided. Moreover, better removal of Cr and other physicochemical parameters has been obtained using 0.2 mm membrane thickness itself.

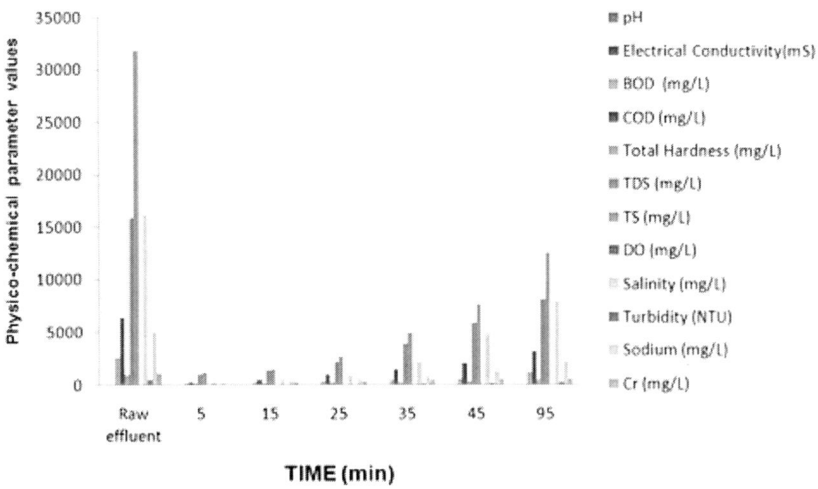

Figure 12. Physicochemical parameters of the treated effluent at thickness = 0.2 mm

From the results, it was clear that the membrane thickness was a function of permeation characteristics and it was proved by Gotoh et al. (1993) that the flux for aqueous solution of dextran sodium sulfate decreased with an increase in membrane thickness, while the rejection for dextran sodium sulfate increased slightly. In another study by Barakat (2008), the metal ions rejection efficiency values were increased with pH increase, reaching maxima values at pH = 7, i.e., 97.6, 99.1, and 99.5 %, for Cu(II), Ni(II), and Cr(III) ions, respectively. Moreover, the percentage of TDS retention was only 86.8 % for 0.1 mm thickness whereas it was 94 % for 0.2 mm thickness which showed a closer value obtained by Kowalik-Klimczak and Gierycz (2014) which is 99.6 % of retention by UF.

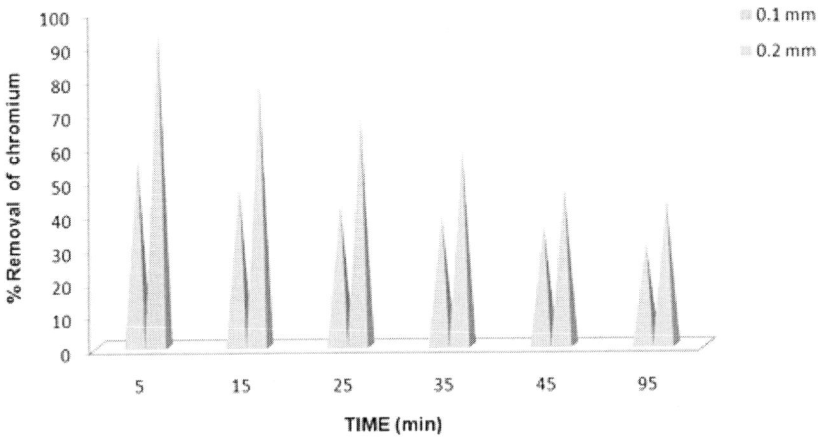

Figure 13. Effect of time on percentage removal of chromium at thicknesses 0.1 and 0.2 mm

3.4.3 Effect of applied pressure

Ultrafiltration process itself is a pressure-driven process except that it needs only minimum pressure for the filtration process to take place. In the porous UF membrane, the transport of solvent is directly proportional to the applied pressure (Trivunac and Stevanovic 2006). Two different pressures namely 50 and 100 kPa have been employed in this study and analyzed the changes that have taken place before and after filtration.

Figure 14 shows the effect of pressure on the removal of various physicochemical parameters at 50 kPa. The EC was found to be reduced from 27.3 to 7.22, 8.42, 9.31, 9.88, 10.74 and 13.5 mS at 5, 15, 25, 35, 45 and 95 min, respectively. The BOD and COD were found to be 2556 and 6419 mg/l which were reduced to 315, 411, 599, 628, 781 and 1301 mg/l for BOD and 525, 731, 1066, 2216, 3022 and 3984 mg/l for COD at 5, 15, 25, 35, 45 and 95 min, respectively. Other parameters were also found to be reduced to a considerable amount. The heavy metal chromium was also reduced to a greater extent namely from 1055 to 333, 389, 473, 593, 684 and 749 mg/l.

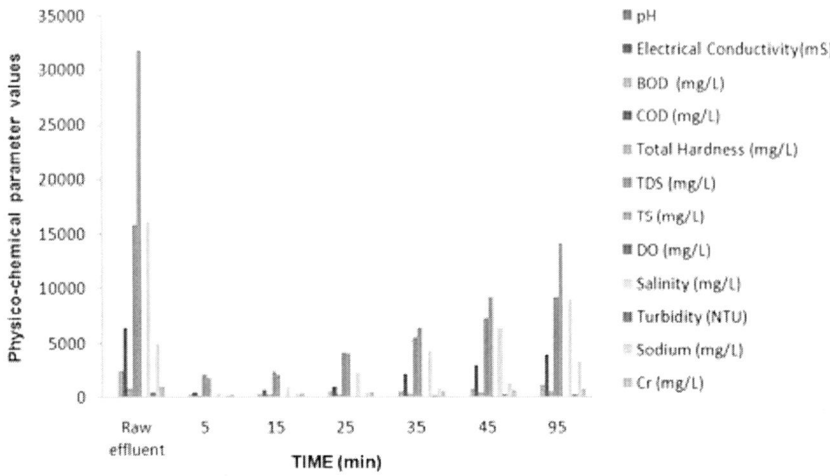

Figure 14. Physicochemical parameters of the treated effluent at pressure = 50 kPa

Figure 15 shows the effect of pressure on the removal of various physicochemical parameters at 100 kPa. All the parameters were found to be greatly enhanced when the pressure was increased from 50 to 100 kPa. As can be seen, the EC values which were 7.22, 8.42, 9.31, 9.88, 10.74 and 13.5 mS using 50 kPa at 5, 15, 25, 35, 45 and 95 min were found to be reduced to 3.22, 5.41, 6.21, 8.16, 10.2 and 13.1 mS at 100 kPa, respectively. The BOD values at 5, 15, 25, 35, 45 and 95 min at 50 kPa were 315, 411, 599, 628, 781 and 1301 mg/l, respectively, which were further decreased to 113, 159, 295, 364, 545 and 1119 mg/l, respectively, at 100 kPa. In the same way, the COD values were also retarded. TH, TDS, TS, salinity, turbidity, and sodium values were all decreased, i.e., more percentage removal was observed when the pressure has been increased from 50 to 100 kPa. In turn, the DO values have been increased for the survival of aquatic organisms.

Figure 16 shows the percentage removal of chromium at pressures 50 and 100 kPa during various time intervals. It was very much evident that the percentage removal of chromium which was only 68.436 % at 50 kPa was improved at 100 kPa whose removal percentage was found to be 95.640 %. Similar trend was observed as the time was increased.

But there is a reduction in percentage removal of chromium with time which may be due to the applied pressure that the pollutants adsorbed in the pore of the membrane move slowly to the permeate during filtration. Thus, the permeate chromate concentration increased with the passage of time (Bade et al. 2008). Thus, the membrane was capable of removing 29.005 % of chromium at 50 kPa and 44.455 % of chromium at 100 kPa at 95 min. Hence, with an increase of the applied pressure, better rejection of the metal ions was obtained. Since high percentage chromium removal of 95.640 % was obtained at 100 kPa, further increase in applied pressure was not carried out in this study.

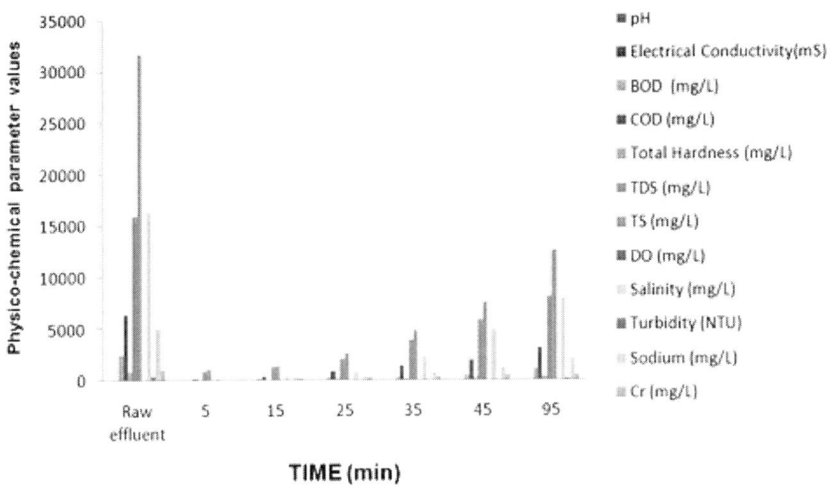

Figure 15. Physicochemical parameters of the treated effluent at pressure = 100 kPa

Also, in a study on the effect of applied pressure (0.4–1 MPa) on percentage solute rejection of chromium and copper ions by Boricha and Murthy (2010), it was found that the rejection increases with increase in feed pressure from 0.4 to 1 MPa for both the ions and the highest rejection was found to be 83.40 and 72.60 % for chromium and copper ions, respectively, at 1 MPa. In addition, the increase in the applied pressure increases the surfactant concentration polarization layer near the membrane surface on which the metal ions gets adsorbed (Kamble and Marathe 2005). Thus, when the charged species flows across a membrane, an unbalanced distribution of the electrical charges at the upstream and downstream interfaces of the membrane takes place. This consequently generates streaming potential, and as stated by Azoug et al. (1997), an increase of the transmembrane pressure may lead to a better rejection of the metal ions.

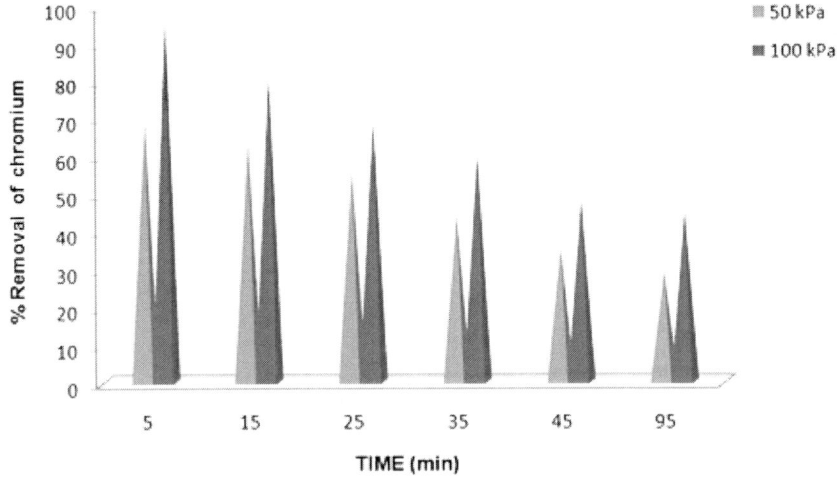

Figure 16. Effect of time on percentage removal of chromium at pressures 50 and 100 kPa

4. CONCLUSIONS

In this study, nanochitosan was prepared successfully with particle size of 100 nm. The membrane was prepared using phase inversion method and characterized using FTIR, XRD and SEM. Ultrafiltration characterization was done by membrane compaction, pure water flux, water content and porosity. FTIR result revealed that the polymers used have been blended well. XRD showed that the membrane is amorphous in nature and SEM exhibited rough surface and large number of pores. The results showed that the membrane obtained a pure water flux of

25.32 l/m^2 h. The water content was found to be 24 %, and a high porosity of 83 % was obtained which showed the membrane's high hydrophilicity and more porous nature. So the prepared membrane was used for the filtration of tannery effluent collected from Ranipet Industrial area. It can be concluded from this study that the membrane CA/NCS/PEG in 1:2:2 ratio was found to be suitable for reducing the physicochemical parameters such as EC, BOD, COD, TH, TDS, TS, salinity, turbidity and sodium as well as the heavy metal chromium from tannery effluent in an effective and efficient way. The effect of pH, membrane thickness and the applied pressure was analyzed and found that maximum removal has taken place when the effluent was in neutral condition, at 0.2 mm membrane thickness and pressure 100 kPa. This membrane ratio could further be implemented at industrial level for the utmost benefit of mankind.

ACKNOWLEDGEMENTS

One of the authors Ms. P. AngelinVinodhini acknowledges with gratitude the financial support (JRF) given by University Grants Commission (UGC), New Delhi. P.N.Sudha acknowledges University Grants Commission (UGC), New Delhi, India, for funding the project.

COMPETING INTERESTS

None of the authors have any competing interests including direct or indirect financial relations with any of the trademarks and company mentioned in this paper except that University Grants Commission (UGC), New Delhi, India, has funded this project.

AUTHORS' CONTRIBUTIONS

PNS and PAV contributed to the conception and design of the study. PAV carried out the experiments, analyzed the data, and drafted the manuscript. PNS supervised the works. Both authors read and approved the final manuscript.

REFERENCES

1. Abdelrasoul, A, Doan, H, Lohi, A, & Cheng, C-H (2015). Morphology control of polysulfone membranes in filtration processes: a critical review. *Chem Bio Eng Rev, 2*(1), 22–43.

2. Ananth, A, Arthanareeswaran, G, & Wang, H (2012). The influence of tetraethylorthosilicate and polyethyleneimine on the performance of polyethersulfone membranes. *Desalination, 287*, 61–70.

3. APHA. (1993). *Standard methods for the examination of water and wastewaters* (p. 874). NW, Washington DC: American Public Health Association.

4. Arthanareeswaran, G, Thanikaivelan, P, Srinivasan, K, Mohan, D, & Rajendran, M (2004). Synthesis, characterization and thermal studies on cellulose acetate membranes with additive. *European Polymer Journal, 40*, 2153–2159.

5. Arthanareeswaran, G, Thanikaivelan, P, Raguime, J A, Raajenthiren, M, & Mohan, D (2007). Metal ion separation and protein removal from aqueous solutions using modified cellulose acetate membranes: role of polymeric additives. *Separation and Purification Technology, 55*, 8–15.

6. Azoug, C, Sadaoui, Z, Charbit, F, & Charbit, G (1997). Removal of cadmium from wastewater by enhanced ultrafiltration using

surfactants. *The Canadian Journal of Chemical Engineering, 75*(4), 743–750.

7. Bade, R, Lee, S H, Jo, S, Lee, H-S, & Lee, S-e (2008). Micellar enhanced ultrafiltration (MEUF) and activated carbon fibre (ACF) hybrid processes for chromate removal from wastewater. *Desalination, 229*, 264–278.

8. Barakat, M A (2008). Removal of Cu(II), Ni(II), and Cr(III) ions from wastewater using complexation-ultrafiltration technique. *Journal of Environmental Science and Technology, 1*(3), 151–156.

9. Barbooti, M M (2015). Simultaneous removal of chromium and lead from water by sorption on Iraqi Montmorillonite. *Journal of Environmental Protection, 6*, 237–249.

10. Boonlertniruni, S, Boobraung, C, & Suvanasara, R (2008). Application of chitosan in rice production. *Journal of Metals Materials and Minerals, 18*, 47–52.

11. Boricha, A G, & Murthy, Z V P (2010). Preparation of N, O-carboxymethyl chitosan/cellulose acetate blend nanofiltration membrane and testing its performance in treating industrial wastewater. *Chemical Engineering Journal, 157*, 393–400.

12. Chen, S-S, Hsu, B-C, Ko, C-H, & Chuang, P-C (2008). Recovery of chromate from spent plating solutions by two-stage nanofiltration processes. *Desalination, 229*, 147–155.

13. Chowdhury, M, Mostafa, M G, Biswas, T K, & Saha, A K (2013). Treatment of leather industrial effluents by filtration and coagulation processes. *Water Resources and Industry, 3*, 11–22.

14. Crini, G (2005). Recent developments in polysaccharide-based materials used as adsorbents in wastewater treatment. *Progress in Polymer Science, 30*, 38–70.

15. De Lima, R, Feitosa, L, Santo Pereira, A E, De Moura, M R, Fauze Ahmad Aouada, F A, Mattoso, L H C, Fraceto, L F (2010). Evaluation of the genotoxicity of chitosan nanoparticles for use in food packaging films.*Journal of Food Science, 75*, 89–96.

16. Duruibe, J O, Ogwuegbu, M O C, & Egwurugwu, J. N. (2007). Heavy metal pollution and human biotoxic effects. *International Journal of Physical Sciences, 2*(5), 112–118.

17. El Hadrami, A, Adam, L R, El Hadrami, I, & Daayf, F (2010). Chitosan in plant protection. *Marine Drugs, 8*, 968–987.

18. Eren, E, Sarihan, A, Eren, B, Gumus, H, & Kocak, F O (2015). Preparation, characterization and performance enhancement of polysulfone ultrafiltration membrane using PBI as hydrophilic modifier. *Journal of Membrane Science, 475*, 1–8.

19. Gökşen, Ç (2005). *Development of a membrane based treatment scheme for water recovery from textile effluents (M.Sc. Thesis)*. Ankara: Middle East Technical University.

20. Gorey, C, & Escobar, I C (2011). N-isopropylacrylamide (NIPAAM) modified cellulose acetate ultrafiltration membranes. *Journal of Membrane Science, 383*, 272–279.

21. Gotoh, M, Tamiya, E, & Karube, I (1993). Preparation and performance of poly (vinyl butyral) membrane for ultrafiltration. *Journal of Applied Polymer Science, 48*, 67–73.

22. Habibi, S, Nematollahzadeh, A, & Mousavi, S A (2015). Nano-scale modification of polysulfone membrane matrix and the surface for the separation of chromium ions from water. *Chemical Engineering Journal, 267*, 306–316.

23. Ham-Pichavant, F, Sebe, G, Pardon, P, & Coma, V (2005). Fat resistance properties of chitosan-based paper packaging for food applications. *Carbohydrate Polymers, 61*, 259–265.

24. Idris, A, & Yet, L K (2006). The effect of different molecular weight PEG additives on cellulose acetate asymmetric dialysis membrane performance. *Journal of Membrane Science, 280*, 920–927.

25. Jayakrishnan, P, & Ramesan, M T (2016). Synthesis, characterization and properties of poly (vinyl alcohol)/ chemically modified and unmodified pumice composites. *Journal of Chemical and Pharmaceutical Sciences, 1*, 97–104.

26. Jyothi, M S, Nayak, V, Padaki, M, Geetha Balakrishna, R, & Ismail, A F (2014). The effect of UV irradiation on PSf/TiO_2 mixed matrix membrane for chromium rejection. *Desalination, 354*, 189–199.

27. Kamble, S B, & Marathe, K V (2005). Micellar-enhanced ultrafiltration of chromate [Cr(VI)] ion from aqueous streams by using cationic surfactant. *Indian Journal of Chemical Technology, 12*, 393–400.

28. Karuppasamy, K, Thanikaikarasan, S, Antony, R, Balakumar, S, & Shajan, X S (2012). Effect of nanochitosan on electrochemical, interfacial and thermal properties of composite solid polymer electrolytes. *Ionics, 18*(8), 737–745.

29. Kato, Y, Onishi, H, & Machida, Y (2003). Application of chitin and chitosan derivatives in the pharmaceutical field. *Current Pharmaceutical Biotechnology, 4*, 303–309.

30. Kesting, R E (1985). *Synthetic polymeric membranes: a structural perspective*. New York: Wiley.

31. Khulbe, K C, Gagne, S, Mohammadi, A T, & Lamarche, A M (1995). Investigation of polymer morphology of integral asymmetric membranes by ESR and raman spectroscopy and its comparison with homogeneous films. *Journal of Membrane Science, 98*, 201.

32. Kowalik-Klimczak, A, & Gierycz, P (2014). Application of pressure membrane processes for the minimization of the

noxiousness of chromium tannery wastewater. *Problemy Eksploatacji – Maintenance Problems, 1*, 71–79.

33. Krishanamoorthi, S, Sivakumar, V, Saravanan, K., & Sriram Prabhu, T V (2009). Treatment and reuse of tannery waste water by embedded system. *Modern Applied Science, 3*(1), 129–134.

34. Kumar, M N V R, Muzzarelli, R A A, Muzzareli, C, Sashiwa, H, & Domb, A J (2004). Chitosan chemistry and pharmaceutical perspectives. *Chemical Reviews, 104*, 6017–6084.

35. Kuttowy, O, & Sourirajan, S (1975). Cellulose acetate ultrafiltration membranes. *Journal of Applied Polymer Science, 19*, 1449–1460.

36. Li, J F, Xu, Z L, Yang, H, Yu, L Y, & Liu, M (2009). Effect of TiO_2 nanoparticles on the surface morphology and performance of microporous PES membrane. *Applied Surface Science, 255*, 4725–4732.

37. Loeb, S, & Sourirajan, S (1963). Sea water demineralization by means of an osmotic membrane. In *Advances in Chemistry* (pp. 117–132). Washington, DC: ACS publications.

38. Maheswari, P, Prasannadevi, D, & Mohan, D (2013). Preparation and performance of silver nanoparticle incorporated polyetherethersulfone nanofiltration membranes. *High Performance Polymers, 25*(2), 174–187.

39. Malik, D, Singh, S, Thakur, J, Singh, R K, Kaur, A, & Nijhawan, S (2014). Heavy metal pollution of the Yamuna river: an introspection. *International Journal of Current Microbiology and Applied Science, 3*(10), 856–863.

40. Mohammadpour Dounighi, N, Behfar, A, Ezabadi, A, Zolfagharian, H, & Heydari, M (2010). Preparation of chitosan nanoparticles containing *Naja naja oxiana* snake venom. *Nanomedicine: Nanotechnology, Biology, and Medicine, 6*, 137–143.

41. Moosa, A A, Ridha, A M, & Abdullha, I N (2015). Chromium ions removal from wastewater using activated Iraqi Bentonite. *International Journal of Innovative Research in Science, Engineering and Technology, 4*(2), 15–25.

42. Muthukrishnan, M, & Guha, B K (2008). Effect of pH on rejection of hexavalent chromium by nanofiltration.*Desalination, 219*, 171–178.

43. Nara, S, & Komiya, T (1983). Studies on the relationship between water-saturated state and crystallinity by the diffraction method for moistened potato starch. *Starch-starke, 35*, 407–410.

44. Nath, K (2008). *Membrane separation processes.* New Delhi: PHI Learning Pvt. Ltd.

45. Ng, L T, & Swami, S (2005). IPNs based on chitosan with NVP and NVP/HEMAsynthesised through photoinitiator-free photopolymerisation technique forbiomedical applications. *Carbohydrate Polymer, 60*, 523–528.

46. Northcott, K A, Snape, I, Scales, P J, & Stevens, G W (2005). Dewatering behaviour of water treatment sludges associated with contaminated site remediation in Antarctica. *Chemical Engineering Science, 60*(24), 6835–6843.

47. Osada, V, & Nakagawa, I (1992). *Membrane science and technology.* New York: Marcel Dekker.

48. Pang, M, Liu, B, Kano, N, & Imaizumi, H (2015). Adsorption of chromium (VI) onto activated carbon modified with $KMnO_4$. *Journal of Chemistry and Chemical Engineering, 9*, 280–287.

49. Pusch, W. & Mossa, G (1978). Influence of pressure and/or pressure differential on membrane permeability.*Desalination, 24*, 39–53.

50. Qu, P, Tang, H, Gao, Y, Zhang, L-p, & Wang, S (2010). Polyethersulfone composite membrane blended with cellulose fibrils. *BioResources, 5*(4), 2323–2336.

51. Rajesh, S, Maheswari, P, Senthilkumar, S, Jayalakshmi, A, & Mohan, D (2011). Preparation and characterisation of poly (amide-imide) incorporated cellulose acetate membranes for polymer enhanced ultrafiltration of metal ions. *Chemical Engineering Journal, 171*, 33–44.

52. Rinoudo, M (2006). Chitin and chitosan: properties and applications. *Progress in Polymer Science, 31*, 603–632.

53. Rowe, R C, Sheskey, P J, & Weller, P J (2003). *Handbook of pharmaceutical excipients* (4th ed.). Washington, DC: American Pharmaceutical Association.

54. Saljoughi, E.,Sadrzadeh, M, & Mohammadi, T (2009). Effect of preparation variables on morphology and pure water permeation flux through asymmetric cellulose acetate membranes. *Journal of Membrane Science, 326*, 627–634.

55. Shi, F, Ma, Y, Ma, J, Wang, P, & Sun, W (2012). Preparation and characterization of PVDF/TiO_2 hybrid membranes with different dosage of nano-TiO_2. *Journal of Membrane Science, 389*, 522–531.

56. Sivakami, M S, Gomathi, T, Venkatesan, J, Jeong, H-S, Kim, S-K, & Sudha, P N (2013). Preparation and characterization of nano chitosan for treatment wastewaters. *International Journal of Biological Macromolecules, 57*, 204–212.

57. Sivakumar, M, Malaisamy, R, Sajitha, C J, Mohan, D, Mohan, V, & Rangarajan, R (2000). Preparation and performance of cellulose acetate-polyurethane blend membranes and their applications—II. *Journal of Membrane Science, 169*, 215–228.

58. Sivakumar, M, Mohan, D R, & Rangarajan, R (2006). Studies on cellulose acetate-polysulfone ultrafiltration membranes II. Effect

of additive concentration. *Journal of Membrane Science, 268*, 208–219.

59. Subbu, C, Chithra Mathew, M, Kesavan, K, & Rajendran, S (2014). Electrochemical, structural and optical studies on poly(vinylidene chloride-co-acrylonitrile) based polymer blend membranes. *International Journal of Electrochemical Science, 9*, 4944–4958.

60. Sudha, P N (2010). Chitin/chitosan and derivatives for wastewater treatment. In S-K Kim (Ed.), *Chitin, Chitosan, Oligosaccharides and Their Derivatives* (pp. 561–585). Boca Raton: CRC Press.

61. Suh, J K F, & Matthew, H W T (2000). Application of chitosan-based polysaccharide biomaterilas in cartilage tissue engineering: a review. *Biomaterials, 21*, 2589–2598.

62. Sun, L, Du, Y, Yang, J H., Shi, X W, Li, J, Wang, X H, Kennedy, JF (2006). Conversion of crystal structure of the chitin to facilitate preparation of a 6-carboxychitin with moisture absorption–retention abilities. *Carbohydrate Polymers, 66*(2), 168–175.

63. Tamura, M, Uragami, T, & Sugihara, M (1981). Studies on synthesis and permeabilities of special polymer membranes, CA30. Ultrafiltration and dialysis characteristics of cellulose nitrate poly (vinylpyrrolidone) polymer blend membranes. *Polymer, 22*, 829–835.

64. Trivedy, R K, & Goel, P K (1986). *Chemical and biological methods for water pollution studies* (pp. 43–45). Karad: Environ. Publications.

65. Trivunac, K, & Stevanovic, S (2006). Removal of heavy metal ions from water bycomplexation-assisted ultrafiltration. *Chemosphere, 64*, 486–491.

66. Tucci, M G, & Rocotti, G (2003). Chitosan and gelatin as engineered dressing for wound repair. *Journal of Bioactive and Compatible Polymer, 16*, 145–157.

67. Vellingiri, K, Ramachandran, T, & Senthilkumar, M (2013). Eco-friendly application of nano chitosan in antimicrobial coatings in the textile industry. *Nanoscience and Nanotechnology, 3*(4), 75–89.

68. Velu, S, Muruganandam, L, & Arthanareeswaran, G (2011). Effect of solvents on performance of polyethersulfone ultrafiltraion membranes for separation of metal ions. *International Journal of Chemical and Analytical Science, 2*, 82–86.

69. Verliefde, A R D, Cornelissen, E R, Heijman, S G J, Verberk, J Q J C, Amy, G L, Van der Bruggen, B, van Dijk, J C (2008). The role of electrostatic interactions on the rejection of organic solutes in aqueous solutions with nanofiltration. *Journal of Membrane Science, 322*, 52–66.

70. Vijayalakshmi, K, Gomathi, T, & Sudha, P N (2014). Preparation and characterization of nanochitosan/sodium alginate/ microcrystalline cellulose beads. *Der Pharmacia Lettre, 6*(4), 65–77.

71. Vimal, S, Abdul Majeed, S, Taju, G, Nambi, K S N, Sundar Raj, N, Madan, N, Farook, M A, Rajkumar, T, Gopinath, D, Sahul Hameed, A S (2013). Chitosan tripolyphosphate (CS/TPP) nanoparticles: preparation, characterization and application for gene delivery in shrimp. *Acta Tropica, 128*, 486–493.

72. Wienk, I M, Boom, R M, Beerlage, M A M, Bulte, A M W, Smolders, C A, & Strathmann, H (1996). Recent advances in the formation of phase inversion membranes made from amorphous or semi-crystalline polymers.*Journal of Membrane Science, 113*, 361–371.

73. Witte, P V, Dijkstra, P J, Vanden Berg, J W A, & Feijen, J (1996). Phase separation process in polymer solution in relation to membrane formation. *Journal of Membrane Science, 117,* 1–31.

74. Yao, Z, Shiyuan, D, Zhang, Y, Zhu, B, Zhu, L, & John, A E (2015). Positively charged membrane for removing low concentration Cr(VI)in ultrafiltration process. *Journal of Water Process Engineering, 8,* 99–107.

75. Yu, J H. Du, Y M, & Zheng, H (1999). Blend films of chitosan—gelation. *Journal Wuhan University, 45,* 440–444.

76. Zeng, J, Ye, H, & Hu, Z J (2009). Application of the hybrid complexation-ultrafiltration process for metal ion removal from aqueous solutions. *Journal of Hazardous Materials, 161*(2-3), 1491–1498.

77. Zeng, G, Lian, G, Zhang, Y, Lu, G, Zhou, Y, Qiu, J, van der Bruggencd, B, Shen, J (2016). Potential applications of abandoned aromatic polyamide reverse osmosis membrane by hypochlorite degradation. *RSC Advances, 6,* 12263–12271.

INDEX